VIDA
ZOO-CIAL

AF794984

VIDA ZOO-CIAL

The Buenos Aires Zoo and the Making of Argentine Society, 1875–1924

ASHLEY ELIZABETH KERR

VANDERBILT UNIVERSITY PRESS
Nashville, Tennessee

Copyright 2025 Vanderbilt University Press
All rights reserved
First printing 2025

Library of Congress Cataloging-in-Publication Data

Names: Kerr, Ashley Elizabeth, 1984- author
Title: Vida zoo-cial : the Buenos Aires Zoo and the making of Argentine society, 1875-1924 / Ashley Elizabeth Kerr.
Description: Nashville, Tennessee : Vanderbilt University Press, 2025. | Includes bibliographical references and index.
Identifiers: LCCN 2025006443 (print) | LCCN 2025006444 (ebook) | ISBN 9780826508065 paperback | ISBN 9780826508072 hardcover | ISBN 9780826508089 epub | ISBN 9780826508096 pdf
Subjects: LCSH: Jardín Zoológico Municipal de Buenos Aires | Zoos--Argentina--Buenos Aires--History--19th century | Zoos--Argentina--Buenos Aires--History--20th century | Zoos--Social aspects--Argentina--Buenos Aires | Buenos Aires (Argentina)--Social conditions
Classification: LCC QL76.5.A72 B8345 2025 (print) | LCC QL76.5.A72 (ebook)
LC record available at https://lccn.loc.gov/2025006443
LC ebook record available at https://lccn.loc.gov/2025006444

Front cover images from the Biodiversity Heritage Library. Snake: Johann Georg Wagler, *Descriptiones et icones amphibiorum* (Sumtibus J.G. Cotttae, 1833); Orangutan: *The Zoology of the Voyage of H.M.S. Samarang, under the command of Captain Sir Edward Belcher, during the years 1843–1846* (Reeve and Benham, 1850); Hippopotamus: Charles d'Orbigny, ed., *Dictionnaire universel d'histoire naturelle*, vol. 1 (Chez les editeurs MM. Renard, Martinet etc., 1847); Giraffes: *Proceedings of the Zoological Society of London*, vol. 1 (1904); Toucan: John Gould, *A Monograph of the Ramphastidae, or Family of Toucans* (Published by the author, [1852]–1854); Elephant: Richard Lydekker and Philip Lutley Sclater, *The Royal Natural History* (Frederick Warne & Co., 1893–1896).

Printed and bound by CPI Group (UK) Ltd, Croydon, CR0 4YY

For Emily

CONTENTS

ACKNOWLEDGMENTS

I cannot thank María José Micale of Ecoparque enough for granting me access to the piles of boxes and files that are the remains of the Buenos Aires Zoo's archive and library. From the first day we met, when she took me on a tour that included the terrace of the Confitería Águila and the basement of the former big cat house (hard hats required!), her support of my project was invaluable. She and the staff of Ecoparque went above and beyond. This book would not exist had they not welcomed me onto the grounds and given me free rein to poke around in the *depósitos*.

I am also indebted to many others in Argentina, including the archivists and librarians at the Biblioteca Nacional Mariano Moreno (especially Laura Braga), Archivo General de la Nación, Facultad de Filosofía y Letras of the University of Buenos Aires, Instituto de Historia Argentina y Americana "Dr. Emilio Ravignani," Biblioteca del Congreso de la Nación, and the Academia Nacional de la Historia de la República Argentina. Mauricio V. Genta was a big help with obtaining high-resolution images. Rodrigo Caresani, José Luis Gamarra, and Vera Senderowicz Guerra provided ideas, texts, and friendship, while Jessica Sarah Rinland inspired me in our conversations about her film *Monólogo colectivo*. Finally, I may never have finished this book without Cole Rizki, who was an invaluable writing partner in person and on Zoom. An enormous thank you to all.

At the University of Idaho, this book was supported by the College of Letters, Arts and Social Sciences, including through summer research and travel grants; Dean Sean Quinlan; the tireless ILL staff; and my colleagues in the School of Global Studies and across the college.

Many people have heard, read, and commented on drafts of this book. Thank you to Oliver Hochadel and the members of his research group for excellent feedback, as well as to colleagues at LASA, SECOLAS, and HSS, and many others for listening to me rattle on about the zoo. Huge thanks to Kristin Haltinner, Rebecca Scofield, and Erin James for their often completely contradictory advice and to Sophie Gilbert for being my "zoology consultant" and provider of disgusting animal facts. Zack Gresham first heard the proposal for this book and believed in it; Steven Rodríguez and Joell Smith-Borne have handily ushered it through the writing and publication process and given excellent advice along the way. My most heartfelt thanks as well to the three anonymous reviewers whose comments made this book many times better.

Jason and Emily were willing to move to Buenos Aires with me for a semester and have always been the best source of inspiration, laughter, and distraction. Thank you to them and my friends and family for keeping me moving forward. Special thanks to Sible, Rowana, and Lucy for putting up with me shooing them off my keyboard and for providing endless purring.

Finally, to all the animals, visitors, and employees of the Buenos Aires Zoo: I hope I've done justice to your stories.

INTRODUCTION

VIDA ZOO-CIAL

The Social Life of the Zoo

A VISIT TO THE BUENOS AIRES ZOO. OCTOBER 20, 1911

A bit jumbled from the crowded tramway ride, you take a deep breath as you walk through the stone gate of the Jardín Zoológico de Buenos Aires, your two small children pulling at your hands impatiently. They have been begging for this visit since your husband showed them the pictures of the newborn camel in this week's issue of *Sherlock Holmes* (much more pleasant subject matter than the magazine's usual gruesome tales of crime). If you are being honest with yourself, you are equally eager to see the adorable baby and get some fresh air.

As you walk further into the park, the street noise and the patriotic proclamations of the beggar leaning near the gate are replaced by animal grunts and children's squeals. You pass a *manicero* and the smells of animals and manure mix with the more enticing scent of toasted peanuts wafting from his train-shaped machine. To your left looms the lion enclosure. With its elaborate ironwork and cupulas, it would not be out of place in Europe or beside the mansions surrounding the zoo. You stop to look at the beasts, but the children are quickly bored. Despite their fierce reputation, the lions lie sluggishly on the concrete floor, dozing in the sunlight like house cats. Not even a girl shoving a stick through the bars and shouting can disturb them.

Everywhere is noise and activity. Chattering monkeys and a lonesome tapir, guards scolding boys for running on the grass. The choo choo of the Lilliputian train doing slow loops around the grounds and the murmur of a teacher reading to her distracted charges from the zoo guidebook. You pass the greenish water of the man-made lake and spy the earthen corral with the sign displaying the camels' scientific name and the map of where they live in the wild. Your children race ahead. "There it is, Mommy!" Balancing on spindly legs, the baby camel hides behind its mother, making space between itself and the visitors banging on the wire fence. "Look what a good mother she is," you say to your daughter, "see how she protects her baby?" Your son pulls on your dress and asks, "Where's the daddy camel?" You look around the enclosure, but before you can respond, a zookeeper with a heavy Italian accent chimes in to explain that Director Onelli moved the father camel to a separate enclosure because it was aggressive toward the baby. It will not be safe to reunite the family until the infant has stopped nursing.

Once you have cooed over the baby, you slowly work your way back across the grounds. At the hippopotamus house, a zoo employee is busy sanitizing the railings while another holds a drooping mass of vegetation over the creature's enormous mouth. You overhear a man sarcastically comment that the hippo eats better than many city residents. The animals do seem to live a luxurious life, hand fed in their palaces. As you leave the zoo, your daughter stops to sound out the words on a metal plaque near the administrative building: "Sea compasivo con los animales" (Be kind to animals). The door opens and out steps the director himself, a mustachioed man you recognize from pictures in *Caras y Caretas*. "That's right," he nods approvingly to your daughter, "the good treatment of animals is a refinement of culture and a preparatory school for being charitable to mankind."[1]

This narrative imagines the experience of a typical middle-class Argentine family's visit to the beloved city zoo in the early twentieth century. Curiosity and recreation motivated my fictional family's trip to the zoo, but once there, they were exposed to messages about hygiene, social class, good motherhood, nuclear families, and morality. This book argues that in the late nineteenth and early twentieth centuries, city elites used the Buenos Aires Zoo and its animals to address social concerns and shape residents into ideal members of *porteño* society and the nation. The city was changing rapidly, and these changes destabilized existing power structures. They also introduced worrisome issues such as poverty and crime, often considered together under the label *la cuestión social*, or "the social question," an umbrella term shot through with racial and class concerns. Founded to "cultivate the spirit of the masses," the Buenos Aires Zoo was an important partner in resolving these issues.[2] Its physical spaces, programming, and representations were pressed into service to educate the masses, especially immigrants and the poor. Ultimately, although these projects aimed to improve the lives of all city residents, they were quite conservative, supporting only superficial reforms while reaffirming established power structures.

This book also advances a secondary thesis that, although the Buenos Aires Zoo and its animals were enticing tools for addressing turn-of-the-century social concerns, their reception and effectiveness was highly variable. As Lisa Uddin has written, "zoos have been recognized as notoriously slippery mediums of communication."[3] Decision makers attempted to exert control over the masses, but zoo visitors had their own experiences of animals and spaces. And while symbolic animals could be wrangled on the pages of a magazine, the real creatures at the zoo were individuals who acted upon their own needs and desires, within the limitations of captivity. The lions who bored children with their indolence and the baby camel who frustrated eager visitors by hiding behind its mother were only some of the beasts who failed to meet visitors' expectations. *Vida Zoo-cial* thus alternates chapters detailing efforts to shape society with Interludes that demonstrate how humans and nonhuman animals challenged, resisted, or appropriated elites' projects for their own purposes. Rather than a peaceful arcadia where people and animals of all types coexisted for the betterment of society, the zoo was a place of constant conflict and negotiations.

The zoo was uniquely well-situated to grapple with the social problems in Buenos Aires. *Vida Zoo-cial* resuscitates this complex participation in shaping the populace, while also highlighting how real-life animals and zoological knowledge undergirded those efforts. At its peak, the zoo had over 1.5 million visitors a year, almost exactly equal to the city's total population.[4] The zoo's low cost to entry, visual nature, broad appeal, and family-friendliness meant that social projects could reach residents of all ages, ethnicities, social classes, and educational levels, while the interpretative openness of zoo animals made them rich in metaphorical potential. It was also a locus for interactions between varied municipal offices and city organizations such as the Asistencia Pública, the charitable Sociedad de Beneficencia, diverse ethnic collectives, the Instituto Bacteriológico Pasteur, multiple hospitals, the University of Buenos Aires, primary and secondary schools, and many others. Furthermore, the physicality of the zoo and its animals makes it easy to perceive the unruliness of subjects of reform and can thus point us to the fractures and resistances present in all attempts to reorder reality. Studying the zoo thus also furthers research on the Argentine *cuestión social* that details the conflicts, contradictions, and alliances drawn between different social sectors and neighborhoods.[5]

Vida Zoo-cial also advances an alternative model for understanding the functions of zoos. Thus far, zoos outside of imperial centers, including those in Latin America, Australia, Africa, and the Middle East, are nearly entirely absent from zoo studies scholarship. When they appear at all, it is as a source of animals, reproducing colonial paradigms that conceive of the regions as providers of raw materials instead of producers of culture. This is the first book-length monograph in English about a Latin American zoo. Studying the Buenos Aires Zoo convincingly demonstrates that zoos in the Global South were not mere imitations of their European and North American counterparts, but rather developed in accordance with local conditions and priorities. Additionally, putting the Argentine experience into conversation with studies of world zoos reveals where the norm of the so-called center might actually be exceptional or odd, as well as points toward new areas of study that were previously eclipsed by dominant interpretative frameworks. Overall, the Argentine experience was more like that of the US than Europe. However, it varied significantly from that of its northern neighbors when political and economic interests collided, especially around pan-Americanism and US influence in the region.

A BRIEF HISTORY OF THE BUENOS AIRES ZOO

Humans have been collecting and displaying animals for thousands of years, for reasons ranging from displays of wealth to scientific study to entertainment. In Europe, the Tiergarten Schönbrunn in Austria (1752) and Ménagerie in the Jardin des Plantes, Paris (1794) lay claim to being the oldest zoos in the world, with many others following in the early nineteenth century, including London (1828), Dublin (1831), Antwerp (1843), and Berlin (1844). The last three decades of the nineteenth century saw an explosion of zoos around the world, including Philadelphia (1874), Cincinnati (1875), Tokyo (1882), Sydney (1883), and Giza (1891). In Latin America, Buenos Aires was the first to open and was contemporaneous with the first US zoos (1875). Río de Janeiro followed in 1888; other countries did not establish their first zoos until the early twentieth century.[6]

The Buenos Aires Zoo came to life as a small section of the newly founded municipal Parque Tres de Febrero, located in the Palermo neighborhood on the northern side of the Argentine capital. Social change and identity shaping were deeply embedded in its roots. As Adrián Gorelik has analyzed, the large park emerged as part of local government efforts to change the city in order to change society; it was a "privileged instrument of reform" promoting social equality and community identity in contrast to the economic interests of the grid.[7] For some urban elites, Parque Tres de Febrero had near mythical potential, and organizers "viewed it as a remedy that had a positive impact on all possible aspects of urban life."[8] Indeed, founder and former national president Domingo F. Sarmiento stated in his speech at the park's opening, "Only in a vast, artistic and accessible Park will the *pueblo* be a *pueblo*: only here will there be no foreigners or nationals, neither oligarchs nor plebeians."[9] It thus was envisioned as a utopian space that could unite city residents and cure social ills.

For the first several years the zoo section was small, largely an afterthought in planning and resource allocation within the larger park system. In 1888, it became a separate institution with its own budget led by Eduardo Ladislao Holmberg, a well-known scientist and doctor.[10] Holmberg was a logical pick. He had founded the nation's first natural history journal, *El naturalista argentino*, with Enrique Lynch Arribálzaga in 1878, and was one of the most important popularizers of Darwinism in the Río de la Plata region.[11] The early years of the nascent zoo focused on building infrastructure and increasing collections. Holmberg oversaw the construction of many of the zoo's iconic buildings, including the principal gates, big cat house, bear palace, monkey houses, and pagoda.[12] He accepted hundreds of donated animals from diplomats, military men, and Buenos Aires's residents, purchasing others, including elephants, polar bears, and a hippopotamus, from famed German animal trader Carl Hagenbeck.[13] Holmberg also used his prior publishing experience to found the *Revista del Jardín Zoológico de Buenos Aires* in 1893. He oversaw three years of monthly publications. The journal situated the zoo as a scientific institution, spread knowledge, and created space for collaboration with other scientists and doctors. Throughout Holmberg's tenure, the Buenos Aires Zoo averaged about 150,000 visitors annually.[14]

Although he was successful in getting the new zoo off the ground and establishing it as place of interest for local scientists and non-scientists alike, Holmberg's projects were hampered by inconsistent funding. He fretted that insufficient cages would lead to tragedy, noting in his 1893 annual report that the polar bear and elephant were immediate threats. As he baldly stated, "we either have to kill them, or we have to build them houses."[15] Holmberg's relationship with the commission overseeing the zoo was often strained, and he was removed from his post as director in late 1903. There are several stories explaining his dismissal. At the time, *Caras y Caretas* suggested that it was because Holmberg refused to bolster the entertainment aspects of the space at the expense of its scientific functions, to make "a circus of novelties" out of "a house of study."[16] Esparrach and Sassaroli hypothesize it had more to do with his Darwinian beliefs or political differences with the mayor of Buenos Aires, Alberto Casares, and Argentine president Julio A. Roca.[17]

Holmberg's replacement was Clemente Onelli, an Italian naturalist who had arrived in Argentina in 1888. In his first fifteen years in the country, Onelli worked at the Museo de la Plata with Francisco P. Moreno, traveled through Patagonia while reporting for the newspaper *El Diario*, helped resolve a border dispute with Chile, and taught natural sciences at the Escuela Normal de la Provincia de Buenos Aires.[18] By all accounts, he was an enthusiastic, educated man with very little direct experience in zoology, but a willingness to learn, speak, and publish on any topic. He invested in expanding the zoo's buildings and artistic offerings, topics well studied by historians of art and architecture like Marina Vasta, Patricia V. Corsani, and Daniel Schávelzon. Onelli also seemed to have no problem balancing the entertainment and scientific aspects of the zoo. He bought, traded for, and/or facilitated the reproduction of several exciting animals, including the first elephant born in captivity in the entire world and a giraffe he paraded through the streets.[19] Onelli also added many attractions for children such as camel and llama rides, a tramway, and a miniature train, and collaborated with educational institutions in the city to hold special events like Arbor Day, Animal Day, and Carnaval celebrations. A master of marketing, Onelli's efforts increased the number of visitors to the zoo from around half a million in 1904 to a peak of over one and a half million in 1913.[20] From then until at least 1921, the zoo would continue to see over a million visitors annually.[21]

Onelli viewed the zoo as a scientific institution despite capitalizing on its ability to entertain. In 1905, he revived Holmberg's *Revista del Jardín Zoológico de Buenos Aires*, publishing extensive observations of the animals' behaviors and psychology in a section called "Idiosincrasias de los pensionistas del Jardín Zoológico de Buenos Aires" (Idiosyncrasies of the pensioners of the Jardín Zoológico de Buenos Aires), as well as the work of local and foreign scientists. He collaborated closely with the physician Dr. Ricardo Lynch to pioneer the field of coprology, the study of feces. He participated in numerous comparative anatomy and pathology projects with Dr. Christofredo Jakob, a neuropsychiatrist at the Hospital de Alienadas. Onelli also developed ties with medical and veterinary schools, ranchers, the Instituto Bacteriológico, and other scientific institutions, giving them access to zoo animals, dead and alive, for research purposes. The knowledge resulting from these partnerships appeared in the *Revista del Jardín Zoológico de Buenos Aires*, *Anales de la Sociedad Científica Argentina*, and international journals. Onelli's correspondence and conference attendance also evidences a worldwide network for science and animal trading.[22] Through these investments and innovations, the Buenos Aires Zoo became a source of national pride. It was an obligatory stop for foreign visitors and many boasted it was equal to European and North American institutions, thus proving the culture and excellence of the Argentine Republic.[23]

VIDA ZOO-CIAL: ZOO AND SOCIETY

The development of the Buenos Aires Zoo occurred against a backdrop of intense economic, social, and political change. During the 1870s, Argentina entered the world market as a supplier of agricultural and raw goods, allowing landholders and oligarchs to amass huge fortunes. Decades of pro-immigration policy and plentiful employment opportunities meant that from 1881 to 1930, 5.8 million foreigners—mostly working-class men—arrived in the port city, speaking a wide variety of languages and looking to try their fortune in the New World.[24] Factories and housing sprung up to accommodate them, while cultural institutions like the new Teatro Colón opera building (1908) proclaimed the nation's modernity. The city center pushed at its limits, gobbling up chunks of the *pampas* to meet demand.[25] From the unrest and disunity of a few decades before emerged a bustling, (relatively) politically stable, modern city that highlighted its progress with the glittery 1910 Centenary celebrations. These included an international exposition, the erection of several of Buenos Aires's most notable monuments, and visits from royalty.[26]

Alongside this glowing facade, however, was a darker reality of poverty, squalid tenement houses, poor working conditions, public health challenges, and crime. The exact dimensions of these social problems, often referred to collectively as *la cuestión social*, were hotly debated across a variety of sectors, from the Catholic Church to the technocratic *higienistas* (hygienists).[27] Many, including the oligarchs of the Partido Autonomista Nacional (PAN) that governed between 1874 and 1916, blamed these issues not on structural inequalities, but on the undesirable characteristics of the growing masses, many of whom were foreign-born. Immigrants were necessary for economic growth but introduced social heterogeneity that could threaten a fragile sense of national unity and Argentine identity.[28] They were also accused of bringing the chaos of anarchism and labor movements to Argentine shores. Furthermore, both immigrants and native-born *obreros* (workers) were diagnosed with biological weaknesses such as susceptibility to alcoholism, sexual deviancy, insanity, and disease. The search was on for strategies to cure the nation's physical and symbolic ills.[29]

Obvious solutions included reforms to secular public education, health inspections, and immigration controls, as well as the development of cultural institutions such as museums and libraries. As this book will demonstrate, the polar bears and lions living behind the massive stone gates of the city zoo were equally important for shaping *porteño* society. A 1908 cartoon in the popular magazine *Caras y Caretas* highlights the zoo's aptness for cultivating the spirit of the masses (fig. 0.1). The drawing depicts two men discussing plans for the Argentine nation. One is fat, in uniform, and carrying a stick, while the other man, skinny with a huge mustache, represents Alfredo Palacios, the first socialist representative in Argentine government. The fat man says to Palacios, "Look, doctor, while a million people have gone to the zoo, almost no one has gone to the library and to the museums. Trust me, when you're a congressperson, make the zoo bigger and eliminate those relics."[30] Although the tone is tongue-in-cheek, the cartoon reflects a fundamental truth: the zoo was by far one of the most popular places in the city. Since 1906, it had received over a million visitors per year; 1,419,679 people passed through its gates in 1911, making it the most visited zoo in the world that year according to director Clemente Onelli.[31]

More importantly, this public was diverse and comprised many of the groups elites most wished to reform. Many of the visitors were children with their parents or participating in early field trips, but the zoo appealed to all ages. As the magazine *PBT* noted, "The zoo, a true children's paradise, where innocence plays among the beasts, is also an escape for adults, attracting more crowds every day."[32] Zoo visitors were also diverse in terms of nationality, gender, occupation, and social class. Ethnic collectives frequently used the grounds for events, plans for the first printed guide included Spanish and Italian editions, and zoo observers frequently noted the number of immigrants that visited. Indeed, during World War I, the zoo experienced a decline in attendance because of the number of Italian reservists who were called back to the homeland to fight.[33] Attendees were also economically diverse, in large part because labor movements, public health specialists, and others successfully advocated for reduced working hours. The 1905 Law 4661, which established Sunday as a day of rest, meant thousands of workers gained more free time.[34] The Buenos Aires Zoo was also relatively affordable, with a ten-cent entrance fee for adults and free entry for soldiers, school children accompanied by a teacher, children under the age of five, and for all visitors on certain Sundays and national holidays.[35] Consequently, the 1907 annual report proudly proclaimed that "Today our Garden has become an integral part of popular customs on holidays and a hygienic outing for other social classes on workdays."[36]

The zoo's great popularity, democratic nature, and non-text-based displays made it particularly useful for reaching the working class and immigrants, hence the suggestion that the government invest in it to improve

FIGURE 0.1. The zoo's popularity made it an effective, modern tool for social change. Mario Zavattaro, "Los hombres prácticos," *Caras y Caretas* 11, no. 532 (December 12, 1908). Image courtesy of University of Miami libraries.

society. Notably, the zoo was a municipal institution, unlike the majority of contemporaneous zoos worldwide that were private or subscription based.[37] The first directors, Eduardo L. Holmberg and Clemente Onelli, were dependent on the municipality for funding and approval of major purchases and construction, and often received direct orders, such as when Onelli was told to open a satellite site in the working-class neighborhood of Parque Patricios.[38] Although the directors had a great deal of influence over the zoo's direction, it is this dependence on the municipality, along with the vast number of zoo stories from diverse authors, that allow us to view the social projects at the zoo as more than a reflection of personal whims. Rather, these projects were negotiated endeavors that integrated the desires and priorities of local government, zoo administration, visitors, teachers, and scientists at institutions across the city.

Existing texts on the Buenos Aires Zoo comprise hagiographic institutional histories including Diego A. del Pino's *Historia del Jardín Zoológico Municipal* (1979) and *Clemente Onelli: El más criollo de los tanos* (2005), and Eduardo Esparrach and Juan Carlos Sassaroli's *Historia del Zoológico de Buenos Aires, 1888–2020* (2022), the previously mentioned studies of its art and architecture, and a few articles, most notable of which is Oliver Hochadel's 2022 study of the Buenos Aires Zoo and global scientific networks. These texts have illuminated aspects of the zoo's institutional history, but scholars have yet to pay sustained attention to its important, multifaceted social functions, its "vida zoo-cial," to borrow the term Onelli used to speak of his animals' social lives.

Similarly, while there are many texts about the early twentieth century in Argentina, particularly on topics such as immigration, motherhood, science, hygiene, the *cuestión social*, popular culture, and labor movements, the zoo appears but rarely. One of the few texts to consider the zoo and the Palermo park in which it was born in the context of identity formation and reform movements is Gorelik's classic text, *La grilla y el parque: Espacio público y cultura urbana en Buenos Aires 1887–1936* (originally published in Spanish in 1998 and translated to English as *The Grid and the Park* in 2022). At its peak, the zoo had over 1.5 million visitors a year, almost exactly equal to the city's total population, and I have found nearly one thousand articles and references to the zoo in newspapers and popular magazines. Given these numbers, it is worth seriously considering the zoo in the context of turn-of-the-century changes and efforts to control or counteract them. Thus, this book introduces new elements to the discussion of these major topics in Argentine history and culture, analyzing where zoo projects and stories supported the narratives emerging from other institutions and where they presented contrasting interpretations of society and the nation's future.

BUILDING A ZOO ARCHIVE

This book is fundamentally an archival project. Thanks to the generosity of María José Micale at Ecoparque Buenos Aires (Buenos Aires Eco-Park, the current iteration of the zoo), I had the opportunity to spend August through December 2022 digging into the zoo's only partially inventoried archive. She gave me space to work on the grounds and allowed me to sift through two rooms of unsorted boxes while Patagonian maras and peacocks wandered around outside the glass doors. In those boxes were the remnants of the zoo's library, as well as correspondence, receipts, municipal reports, and employment records from about 1900 until the 1990s. While many of the earliest documents were in poor shape, they allowed me to build on existing historical narratives to piece together a detailed account of the zoo's early history, particularly regarding social reform. The zoo's own journal also was a rich source, especially for the earliest years, for which municipal and institutional records are fragmentary.

The second piece of my archival research was building a corpus of zoo stories, to use Randy Malamud's "catchall reference to a diverse range of literary and popular cultural artifacts" featuring zoo animals.[39] The Buenos Aires Zoo came of age at the same time as a democratizing print culture, as public education reduced illiteracy from a peak of 78 percent to 35 percent by 1914, the population boomed, the number of Argentines in the middle class grew, and technical innovations reduced printing costs and allowed for the incorporation of photographs.[40] Because of the simultaneous nature of these changes and the zoo's rise to widespread popularity, by the early twentieth century zoo stories were ubiquitous. This abundance of materials allows us to answer Takashi Ito's call to think more critically about how people responded to zoos, either in their correspondence, media coverage, or diaries.[41]

I collected these stories and images of the zoo through archival research at the Biblioteca Nacional Mariano Moreno, Archivo General de la Nación, Instituto de Historia Argentina y Americana "Dr. Emilio Ravignani," and the Biblioteca del Congreso de la Nación. Digital repositories, including HathiTrust, the Academia Nacional de la Historia de la República Argentina, Ibero-Amerikanisches Institut, Biblioteca Digital AECID, British Library's Endangered Archives Programme, and the Biblioteca Nacional de Maestras y Maestros, turned up many other sources. These include technical and scientific journals, conference proceedings, field trip reports, articles from newspapers such as *La Nación* and *La Prensa*, poems, novels, theater plays, photographs, videos, and articles and cartoons from popular magazines such as *Caras y Caretas*, *PBT*, *Billiken*, *Mundo Argentino*, *Fray Mocho*, and *El Hogar*. I also searched online marketplaces and used bookstores for postcards and photographs of the zoo. In total, I found hundreds of references to the zoo.

A brief example of the relationship between the zoo and the rapidly growing popular press, and one of the most important sources for this book, is *Caras y Caretas*, an illustrated weekly magazine whose trajectory echoes that of the zoo. *Caras y Caretas* was founded in 1898, a decade after the zoo, but both peaked in the first two decades of the twentieth century. Like the zoo, *Caras y Caretas* was relatively accessible and had material for the whole family.[42] Its image-heavy design was perfect for promoting the zoo, while the zoo provided ample and attractive material for the magazine's writers, photographers, and illustrators. Some articles were reprints of texts that had appeared in the *Revista del Jardín Zoológico*, others were stories and photographs from the magazine's regular contributors. Even famed Uruguayan author Horacio Quiroga was tasked with writing about the institution. He wrote to a friend that *Caras y Caretas* had begged him for more material, including "short notes for local color [*ilustración callejera*], such as hypnotism, Zoo curiosities, etc."[43] The zoo sold magazines, while magazines such as *Caras y Caretas* promoted zoo attendance, disseminated scientific information, and shaped the wider public's interpretations of the institution's animals and enclosures.

Of course, the Argentine media of the time was diverse and not all outlets had the same relationship to the zoo. *Caras y Caretas* often uncritically repeated the institution's claims and messaging, but other publications such as the socialist paper *La Vanguardia* oscillated between support and criticism. Political ideologies, audiences, thematic focuses, and sales pressures also shaped how Argentine media interpreted zoo animals and events in connection with social issues. Throughout this book, I note moments of critique and dissent, while also affirming that there was a startling degree of cohesion in representations of the zoo and its animals. While some of the texts I study have been incorporated into existing zoo studies, my corpus is many times larger and goes beyond the most popular papers and magazines. Additionally, as a scholar with a background in literary studies, I approach these texts as more than evidentiary proof of the development of the zoo. They were also primary texts that created worlds in addition to reflecting them.

I balance these close readings with a broad selection of secondary literature. I have never met a rabbit hole that I did not immediately go down; this is especially true in this book, which pushes beyond the boundaries of my previous expertise. Of special importance is reading across languages, specifically those works published by Latin American scholars in Spanish that have often been ignored by scholars publishing in the Global North. The result of this research is a complex tapestry of history and stories about the Buenos Aires Zoo from its founding in 1875 until Clemente Onelli's death in 1924. Because of the availability of records and the growing wealth and reach of the zoo, I focus more on the twenty years under Onelli than the period before it was an independent institution (1875–1888) or the sixteen years under Holmberg (1888–1904). Indeed, Onelli himself complained that there were no records prior to when he became director in 1904.[44] However, it is important to include these earlier periods given that the roots of the zoo's heyday in the 1910s began there. I end with Onelli's death as it marked the end of a period of great popularity and successes for the zoo. In the next few years, articles about the zoo in the popular press diminished greatly, and those that did appear turned from highlighting the importance of the zoo to documenting poor conditions and lamenting its fall from grace. This time frame also overlaps neatly with a period of marked changes in Argentine demographics, politics, and economics.

Dogging this text, as with all those about zoos, is the question of animals and their representation. Early zoo studies treated animals as little more than objects controlled and manipulated by humans. More recent texts have highlighted the metaphorical value of animals. In Erica Fudge's words, "we [humans] need animals to think with, live with, observe or we disappear."[45] Animals become highly political, allowing us to draw lines between lives worth protecting and lives to be abandoned.[46] From this perspective, the culturally constructed animals that humans create "have only a tangential relationship with the animals in and of themselves."[47] A third, more radical approach aims to decenter humans entirely, rejecting anthropomorphism and recuperating the stories of the "real" and "individual" animals. These posthumanist scholars underscore the need to turn to alternative (often material) archives and shift entire paradigms to capture a fuller vision of animals' experiences and roles in society.[48]

This book will engage, to varying degrees, with each of these three approaches to understanding animals. At the heart of it, I am interested in how people use the zoo and its animals to shape society, and thus this is not an animal-forward history of the zoo nor one of Éric Baratay's speculative animal biographies.[49] I look at how animals—both in their physical bodies and in their cultural, metaphorical constructions—furthered human projects and human actions. By this, I mean that goats were objects used to teach hygiene and improve infant health, while also symbolizing the proletariat themselves in political cartoons. At the same time, I am firm in my belief that while elites (and humans in general) could make their plans, this did not mean that the human or nonhuman animals involved would react and behave as they wished. My hope, therefore, is to present humans and nonhuman animals at the Buenos Aires Zoo as inextricably intertwined, both in daily existence and in projects of meaning making, in a way that focuses on the human without losing sight of the fact that zoo animals were individuals with their own needs, desires, and perspectives. As in Few and Tortorici's *Centering Animals in Latin American History*, "real animals figure alongside the metaphorical animals that were used to symbolically represent the limits of humanity and thus sometimes occluded the real, sentient animal. Animals appear to be both everywhere and nowhere at the same time."[50]

Vida Zoo-cial joins a rapidly expanding body of zoo literature. Many existing texts are best described as institutional histories, often funded by zoos or zoological societies themselves, and thus highly laudatory.[51] Others have examined the ethics of animal captivity, arguing for the abolishment or serious rethinking of zoological spaces. A third strand of inquiry ties zoos to their scientific functions, examining their value for natural history education, scientific discovery, and conservation movements.[52] Of most interest for this book are those scholars who have teased out the relationships between zoos, zoo animals, and society across time and place.[53] Many of these texts look at how zoos mediate the relationships between nature and culture, city and country, wild and domestic. They also analyze the zoo as both tool and reflection of empire, a relationship summarized by Maneesha Deckha: "The zoo came to be seen as part of the repertoire marking cities as places of high culture and their inhabitants as paradigmatic humans compared to the natural realm of the colonies and hinterlands and their inhabitants."[54]

This book counterbalances those studies, for despite their theoretical and empirical richness, they have largely been geographically constrained, focusing on zoos in the Global North (especially the United States, Canada, Germany, England, and France). Emblematic of this tendency is Tracy McDonald and Daniel Vandersommers's otherwise excellent edited collection, *Zoo Studies: A New Humanities* (2019), in which eight of thirteen chapters focus on the US and Canada, four on European zoos, and one on Japan.[55] This geographically narrow viewpoint limits our interpretative horizons. By turning to Global South zoos, we can begin to ask what happens when empire is no longer the primary frame of analysis; what it means when charismatic megafauna are also common "local" species; and how animal meaning and zoo functions shift across cultures, political systems, climates, linguistic systems, and more. In Latin America, a few scholars have taken up this mantle, most notably Regina Horta Duarte, who published an article on nation building and the Chapultepec Zoo in Mexico as well as an Oxford Research Encyclopedia entry on zoos in Latin America. Overall, however, as Horta Duarte concludes, "The history of zoos in Latin America is yet to be constructed."[56] *Vida Zoo-cial* aims to lay a cornerstone for the future.

A GUIDE TO *VIDA ZOO-CIAL*

Vida Zoo-cial is composed of four primary chapters, each of which looks at how Argentine elites and cultural commentators used zoo spaces, animals, and stories to attempt to address some aspect of the *cuestión social*. Each chapter takes up a major issue of the period: hygiene (public health), class relations and labor, immigration and race, and gender. These divisions, of course, are largely artificial, for the topics are inextricably bound together. How can we speak of labor without speaking of immigration? Or maternalism without considering hygiene? They also intersect with other contemporary concerns. I make occasional references to topics such as educational reform, political elections, animal rights activism, the development of agriculture, Argentine understandings of transformism, and so on, but space and coherence mean that these important topics remain largely outside the scope of this book. Future work is bound to pick up those threads and make excellent contributions.

In Chapter 1, I trace efforts to improve Buenos Aires society through hygiene and public health. The idea of zoos bringing orderly, hygienic recreation to the masses was not unique to Argentina; in Europe and the United States, reformers sought to provide alternatives to bars and brothels for working-class families, improving their moral and physical state.[57] The Buenos Aires Zoo's attempts, however, responded to the contours of its particular historical-political context, namely the influx of immigrants and modernization of the city. The zoo was an educational green space that could counteract the perceived physical and moral decay caused by the living and working conditions of the city's poor. The zoo's cleaning practices and guidebooks also presented visitors with information about how to live hygienically, from antiseptics to vaccination, while satellite sites like model milking parlors and a poultry husbandry school took the latest practical science directly to the populations elites most wanted to improve, populations defined by race and class. And out of sight of the public, zoo employees collaborated with technocrats and scientists at the city's institutions of social control: hospitals, laboratories, and prisons. In all these interactions, animal bodies were essential inputs for improving human health and order.

Chapter 2 turns to class divisions and the political programs designed to overcome, mitigate, or entrench them. In the early twentieth century, Argentine authors developed the image of an aristocratic zoo populated by animals living in luxury and contrasted with "proletariat" animals like horses and donkeys. Although this image was consistent across genres and political ideologies, it (and the physical zoo itself) were put to very different uses by diverse social groups. Some supported emerging labor movements, while others engaged the same tropes to defend the status quo. I also recover the zoo's largely forgotten workforce, piecing together their stories from archival records and elucidating the correspondences between zoo employees and zoo animals. While zoo spaces and images could be used to advocate for the working class, the institution's structure and directors' ideologies curtailed the nature and reach of these proposals.

The *cuestión social* was also fundamentally about immigration. Like the city's humans, zoo animals were categorized as "immigrants" or "creoles," with each category assigned certain values. In Chapter 3, I argue that debates over the ideal composition of the zoo's collections mirrored concerns about the city's increasing cosmopolitanism, as did acclimatization experiments that aimed to adapt exotic species to the local climate. The recommendations that emerged from these metaphorical spaces took concrete form at the zoo as the layout of the grounds promoted assimilation and an identification that was simultaneously municipal and national. Finally, I argue that the Buenos Aires Zoo flips established understandings of zoo and empire on their head: rather than represent empire, the Buenos Aires Zoo and its animals were tools in an anti-imperial, anti-United States defense of the *raza latina*, or Latin race. Throughout these projects, Argentina's Indigenous populations were continually ignored.

Finally, turn-of-the-century Argentine social concerns were not static, but diachronic, concerned fundamentally with progress and thus with future generations. Chapter 4 looks at animal reproduction at the zoo as a stand in for human reproduction, and the zoo itself as a model for generative nation building. From courtship to parenting, the zoo animals provided object lessons in gender roles, reifying a maternalist vision that was sweeping the city. The zoo grounds also provided the necessary conditions for embodying good motherhood, featuring an aseptic nursing salon for promoting breastfeeding and participating in programming designed to reduce infant mortality like the 1917 Semana del Nene (Children's Week). These positive images were juxtaposed with the negative example of the *ñandú* (rhea or American ostrich), a bird whose reproductive behavior gave it ample relevance as a symbol of the most "pernicious" aspects of contemporary feminism. Together, they constructed a relatively conservative view of women's role in society.

In each of these chapters, I detail elite goals, plans, utopias, dreams, and ideas for the institution. However, as Elizabeth Hanson writes in the context of the United States, "zoo visitors experienced the displays in ways that managers did not anticipate, and they did not necessarily follow instructions."[58] Eduardo Ladislao Holmberg and Clemente Onelli, as well as their bevy of supporters, overseers, and employees, could attempt to impart all sorts of social controls, designing zoo grounds and enclosures, publishing articles, and speaking on various topics. But they had no way of ensuring that zoo visitors and the *porteño* public at large would perceive those messages as desired, never mind participate actively in their projects. This book thus also draws attention to their half-successes and failures.

Similarly, while caged zoo animals were dependent on humans for food and other basic needs, the zoo's control over its animals was always partial. Human projects were contingent on animals appearing or behaving in certain ways and the animals themselves often did not lend their bodies to these proposals. They escaped, attacked, refused to reproduce, and died. Animal agency is a contested subject. Some scholars believe assigning agency to animals recognizes their important roles, knocking humans off their pedestal of exceptionalism. For example, Jason Hribal argues that when captive animals resist, "These animals . . . are rebelling with knowledge and purpose. They have a conception of freedom and a desire for it. They have agency."[59] Other scholars, however, note that even "To simply declare that such animals have 'agency' might be to approach the issues of animal sentience and instinct in an anthropocentric way, especially given that the notion of agency arises in a secular Western culture and is itself saturated with Western humanist beliefs."[60] My understanding of animal resistance and agency falls somewhere in between, hewing most closely to Erica Fudge's: animals may not have subjectivity or thought processes like ours, but that does not mean they lack the ability to cause change, affecting their worlds and ours.[61]

To reflect these challenges and realities, each chapter in *Vida Zoo-cial* is followed by a short Interlude that examines moments when humans or animals resisted, challenged, thwarted, or manipulated those projects to suit their own needs. They are far from comprehensive, but rather glimpses of hundreds of similar moments lost in the archival gaps encompassing other species and marginalized human groups. They are meant to be read in dialogue with the main chapters, punching holes in the narratives of elites' plans. The first Interlude examines how animals' and visitors' behaviors contradicted the zoo's attempts to implement hygienic measures, repurposing materials for their own interests and fulfilling biological imperatives. The second points out the idealized fantasies driving classed representations of zoo animals and asks what would happen if we grounded these metaphors in concrete reality and considered animals workers. In the third Interlude, I examine the case of Petronio, a male orangutan who, it was claimed, died by suicide. What happens when animals cannot adapt and do not voluntarily lend their bodies to our projects? The final Interlude looks at how animal sex of all sorts undermines the project of building a heterosexual nation expressed in other areas of the zoo. The Interludes emphasize the many competing interests at the zoo and how elite desires were not a guaranteed outcome. They also serve to recognize nonhuman animals' involuntary participation in these processes. The animals at the Buenos Aires Zoo were living, breathing individuals who, while never perfectly visible to us, also experienced these projects and were affected by them. To ignore them is to leave out half the story.

CHAPTER 1

A LEARNED AND HYGIENIC EXCURSION

Public Health, Medicine, and the Zoo

How does a kangaroo take a bath? A 1922 cartoon titled "La higiene en el zoo" answered that question for the young readers of *Billiken* magazine (fig. 1.1). In it, a happy kangaroo uses its paws to wipe its face, stomach, back, and ears, and the caption explains that "the kangaroo has faith in hygiene carried out without water." In the same image, a panther and a pelican engage in a "hygiene contest," fastidiously cleaning each part of their bodies in species-appropriate ways. A pig refuses to wash, and a horse rolls on the ground taking a dust bath while a duck looks on in horror.[1] Ostensibly a fun and even educational image for children, the 1922 cartoon can also be read as a reflection of nearly forty years of intense focus on hygiene at the Buenos Aires Zoo. From its founding, administrators and city officials made the cleanliness of zoo animals and human visitors a central mission. In subsequent decades, the institution became an increasingly active participant in efforts to solve the city's social problems through hygiene. New programming, the circulation of hygienic zoo stories, and scientific experimentation attempted to improve the mental and physical health of visitors, while the opening of satellite sites closer to working-class and immigrant populations brought these innovations to the masses. Whether for kangaroos or factory workers, the clear message was that hygiene would ensure the future health, unity, and prosperity of the nation.

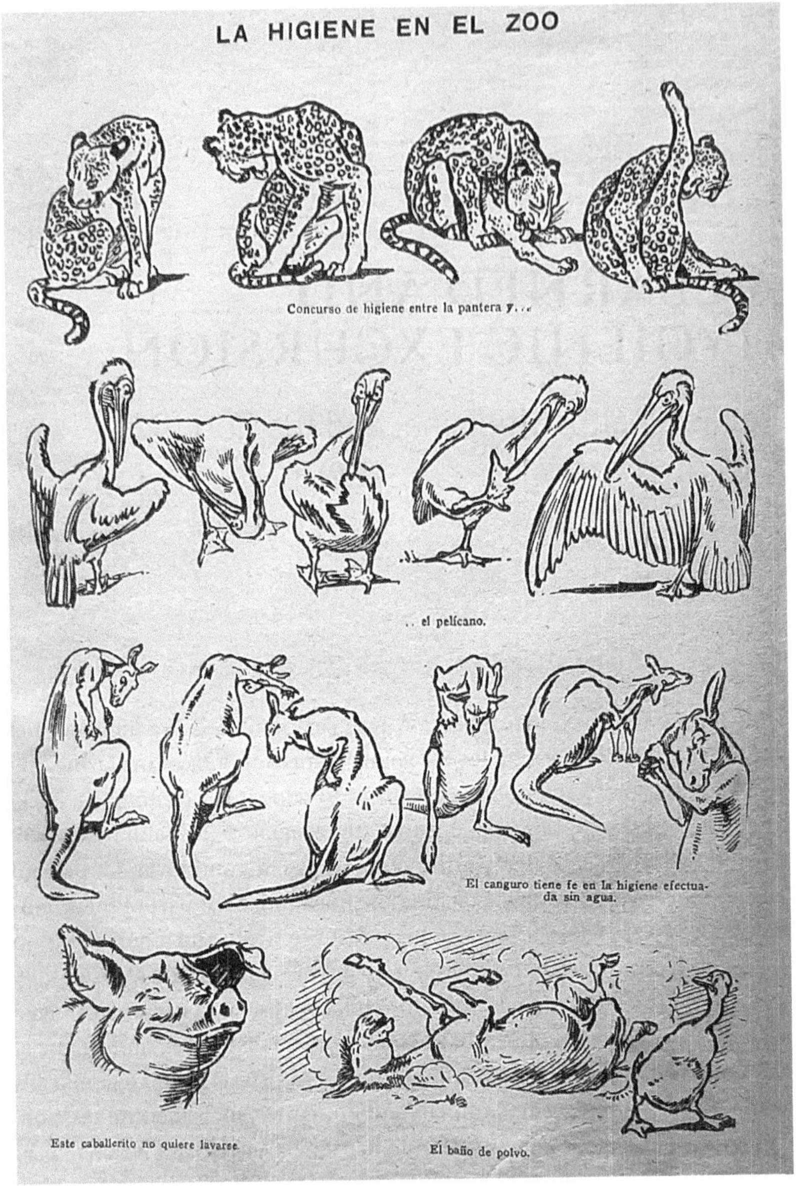

FIGURE 1.1. Even zoo animals knew the importance of staying clean. "La higiene en el zoo," *Billiken* 3, no. 143 (August 14, 1922).

In this chapter, I first define what was meant by a "learned and hygienic excursion," identifying key areas where lawmakers, intellectual elites, and the zoo's administration promoted and taught cleanliness to visitors, particularly those from the working class. I then highlight the close ties between the zoo and Buenos Aires's hygienists and medical practitioners. Zoo animals benefited from many of the hygienic practices positivist reformers were spreading among the human residents of Buenos Aires, improving their comfort and extending their lives. At the same time, their bodies—in life and after death—were necessary for the development of medicine in a way that was not replicated in reverse, at least not among "normal" human patients. Finally, I examine the limitations of the Palermo site for reaching those most in need of hygienic recreation and instruction and consequent efforts to take hygiene directly to the masses. In this chapter I consider hygiene somewhat independently, but later chapters will show that it was not conceptualized nor applied evenly across the city's population; its manifestations cannot be separated from analysis of topics such as labor movements, immigration, nationalism, and gender.

"A LEARNED AND HYGIENIC EXCURSION"

Across the Americas, the end of the nineteenth century and beginning of the twentieth saw rising interest in using the tools of science and hygiene to enact social reform.[2] Buenos Aires was no exception. During this period, population growth and industrialization led to new challenges. Lawmakers and intellectuals looked at the chaos around them, noting poverty, prostitution, and crime. They attributed these social ills to a general national unhealthiness caused by poor living and working conditions, overcrowding, alcohol, and illness, including tuberculosis and syphilis. Perhaps more frightening, physical problems appeared to go hand in hand with moral decay.[3] These concerns were racialized through frequent association with the city's growing immigrant population. In sum, rather than the hoped-for modernization, elites saw themselves surrounded by biological and moral degeneration, reduced productivity, and disunity stemming from a crumbling national identity.

Educated in positivism and a largely neo-Lamarckian understanding of heredity, they turned to science and medicine to counteract these forces and impose the order they so desperately sought. In Julia Rodríguez's words, the *higienistas* firmly believed in "medical health as a forerunner to social and political health" and took definite steps to achieve it.[4] The only way to ensure a strong national race and economic prosperity in the future was to identify the causes of illness and eliminate them. Notably, these efforts were focused on the working class and immigrants.[5] Not all ethnic groups were treated the same: for example, Galicians were associated with higher tuberculosis rates, while most Jews and "indios" were seen as predisposed to disease in general. Explanations of these variances were diverse, including heredity, cultural reasons, and the types of jobs ethnic groups were most likely to hold in the city, and only sometimes overlapped with reality.[6] For example, Armus convincingly shows that historical data does not support the thesis that Jews in Buenos Aires were congenitally weaker; antisemitism, however, does.[7]

Hygienic measures generally emphasized prevention over cures. Many were relatively benign, such as teaching children about hand washing, improving occupational safety, and regulating the food supply. Others, however, opened previously private spaces to intrusive State intervention. These included forced quarantine and/or institutionalization and "an obsessive control, especially of immigrants' bodies, characterized by humiliating removals from apartments in 'unhealthy' neighborhoods, invasive physical exams, and the selective use of fingerprinting."[8] Hygiene, or a perceived lack thereof, thus became an important tool for determining the limits of the increasingly multi-ethnic nation. Finally, concerns over hygiene cut across ideological boundaries, spreading among socialists, oligarchs, feminists, and educators.[9] As Armus demonstrates in *The Ailing City*, these groups articulated varying programs and pathways to health, but there was a broad consensus that the "secular catechism of hygiene" was a cornerstone of achieving modernity.[10]

The Buenos Aires Zoo's founding and rise to peak popularity occurred parallel to the elevation of science and hygiene as the ultimate tools for city administrators and politicians looking to transform the populace. It is therefore unsurprising that, like other municipal institutions of the time, the zoo's founding was at least partially justified by the need to bolster public health. The institution's earliest regulations established in chapter I, article 2 that, "given its nature, the space it occupies, and the distribution of its contents, its object is to provide a hygienic center of solace and instruction to those who visit it."[11] As this sentence makes clear, cleanliness and disease prevention were the necessary foundation for all the hopes that lawmakers had pinned on the zoo. Exotic animals and pony rides would attract and entertain visitors and the growing zoological sciences would have ample subjects to study, but these efforts could only serve the nation if they took place in a hygienic location that would counteract the negative forces of the modern city.

Accordingly, some of the zoo's first years unfolded under the watchful eye of one of the most important hygienists of the era, José Penna. In 1903, Penna was appointed president of the zoo's Comisión Consultiva (advisory commission), where his responsibilities included approving all purchases, hires, programming, and other activities. At the time of his appointment, he was also the director of the Casa de Aislamiento, a quarantine center, and the author of multiple works on cholera and other transmissible diseases. Later, he would be appointed director of the influential Departamento Nacional de Higiene (National Department of Hygiene).[12] Penna's participation surely kept hygiene at the heart of the zoo's activities during critical years for its development.

The zoo's first two directors, Eduardo L. Holmberg and Clemente Onelli, also repeatedly centered hygiene in their reports and correspondence. In 1894, Holmberg emphasized the zoo's "scientific and hygienic double character," situating it as a productive and useful space.[13] As the second director from 1904 to 1924, Onelli made many changes to the zoo, particularly adding commercial attractions such as a miniature train and pony rides that catered more explicitly to young visitors. Nonetheless, he would echo Holmberg's declaration nearly three decades later, calling the institution he presided over a "learned and hygienic excursion."[14] While the relative balance of the zoo's other goals had changed over that thirty-year span, hygiene remained a constant priority.

What did it mean to be a hygienic excursion? For many residents of Buenos Aires, life was experienced in dirty, cramped quarters. The working class grew significantly from the late nineteenth century through 1914.[15] Many worked long hours with limited health and safety protocols. Fire, poor air quality, and accidents were constant threats to the men, women, and children who worked in the city's growing industries.[16] And with few recreational spaces open to impoverished residents, workers passed their time in bars, gambling houses, brothels, and other locations that were perceived by elites to threaten their physical and moral health.

Conditions were often no better at home. Between 1885 and 1920, 10 to 25 percent of the population lived in *conventillos*: crowded, poorly ventilated multifamily homes, frequently lacking basic sanitation.[17] Other forms of precarious housing proliferated, and even single-family working-class homes were often unhygienic.[18] Writer Santiago Estrada pulled no punches in describing city housing, calling the *conventillo* an "epidemic workshop in which each of its filthy beds is the marriage bed where yellow fever and cholera reproduce."[19] Tuberculosis was another major concern, accounting for nearly half of infectious-disease deaths in 1900.[20] Indeed, from the 1890s until the 1930s, the death rate from tuberculosis remained consistently above 170 deaths per 100,000 people.[21] Even the nonhuman animal world criticized Buenos Aires's unhygienic living and working conditions: in an "interview" with the press, a visiting chimpanzee, His Majesty Cónsul III, complained of Buenos Aires, "People work too much, since they live looking for ways to do nothing. The buildings have roofs (this system is inimical to ventilation)."[22] Even an ape recognized the dangers of modern *porteño* life and, it is assumed, lived better than many of the denizens of the city he was touring.

To combat the illnesses of urban life, hygienists prescribed green spaces, fresh air, and moderate exercise. By keeping workers away from unhealthy activities and exposing them to positive forces like fresh air and exercise, both the current generation and their progeny would become stronger and more productive. Among the spreading grid of the city, the park emerged as the "natural antidote to the overcrowding of the poor city and the unhealthiness of factory work," as well as a space for edifying morality and constructing community.[23] Domingo F. Sarmiento's Parque Tres de Febrero in Palermo was the first and perhaps best example, planned and projected as the heart of a new city. The zoo's location in the park made it a natural extension of this project. Years after its founding, director Clemente Onelli advocated for municipal offices to have days off so that workers could get "the necessary rest and the best lung exercise in open air," with the implication that such activities would take place at his institution.[24] Popular magazine *Caras y Caretas* also noted that on holidays, the working class would go to the zoo and other green spaces "so that their children can store oxygen, and tomorrow have strong lungs for work."[25] These quotes make abundantly clear that the zoo's hygienic benefits were not just about individual health and happiness, but also society's future productivity and benefit for the nation.

Reformist educators seized on these benefits and the zoo quickly became one of the most popular destinations for the emerging field trip movement. In 1908, *El Monitor de Educación Común* highlighted the dark green of the trees and grass at the zoo, as well as the white of the paths, images of vitality and purity.[26] In contrast to the gray of the city, the zoo was a healthy environment where "the soul of the future" could learn and develop.[27] At the zoo, teachers and administrators observed how children ran about "breathing the pure and life-giving air deeply," again establishing the zoo grounds as uncontaminated and energizing, in direct contrast to the debilitating squalor of the grid.[28] Field trips to the zoo thus satisfied educational impulses in two ways: first, through the concrete knowledge children acquired through empirical observation of the animals and signage, and second, because exercise, fresh air, and light would stimulate their minds and bodies, preparing them to work harder and think more clearly. As in the quotes from Onelli and *Caras y Caretas* above, educators also explicitly acknowledged the ties between health and productivity and identified the zoo as an ideal locus for mobilizing this relationship in their young charges.

While the zoo was to some extent inherently healthy due to its location, employees had to work hard to maintain cleanliness despite the animal residents who shed, pooped, and created food waste, and the thousands of daily visitors (who also created waste). Throughout the period under study, some of the materials zoo administrators most frequently requested from the municipality were cleaning supplies. In February 1918, employees used sixty-nine straw brooms and four gallons of disinfectant to sanitize human and animal spaces.[29] In May 1919, Onelli requested sixty large straw brushes and three hundred brooms to be used in the second half of the year.[30] In multiple letters, he worked to convince the municipality that it was economic to buy the more expensive, US-manufactured disinfectant Coro Neleum, because it was more concentrated than other products. However, he concluded, he would not be picky because he had been trying for four months to get the municipality to send anything at all.[31]

Employees also worked hard to publicize the institution as a hygienic oasis. One of the first sections of the 1904 Zoo Guide—distributed to visitors at no cost and shared with local schools as educational material—was titled "Comfort, Filtered Water, and Hygiene."[32] This section explained that the water at the zoo was pure and potable, and that fences and rails were disinfected twice a day to counteract children's bad habit of putting their mouths on them. Behind-the-scenes areas were also carefully controlled, with specific procedures for cages and the space where employees sacrificed the animals that would be fed to the zoo's carnivores. Onelli boasted that doctors that came to the Segundo Congreso Médico Latinoamericano's (Second Latin American Medical Conference's) International Exposition of Hygiene in April 1904 declared the latter room so clean that it could be compared to "an aseptic operating room in a hospital."[33] An image labeled "Autopsy Room" similarly shows walls lined in easy-to-clean white tiles, windows for ventilation, and a ceramic autopsy table with a drain, bringing the latest in hygienic design to zoo grounds (fig. 1.2).

In 1906, Onelli insisted that hygiene was "an essential duty" of the zoo, and, paralleling the State's move toward more intrusive hygiene measures, argued for an active control of visitors, including banning those with skin conditions, even if they were beginning to heal. With these measures, he concluded that "we are sure that there is no similar establishment where preventive measures are followed with such exaggerated rigor, but because

FIGURE 1.2. The zoo's autopsy room around 1910. Inventory 221796, box 57, Repositorio Gráfico, Acervo Gráfico Audiovisual y Sonoro, Archivo General de la Nación. Image courtesy of the Archivo General de la Nación.

of them we have had the daily satisfaction of seeing children in the Garden [Zoo] whose families would not have brought them had they not known of these aseptic precautions."[34] While the truth of his statement may be exaggerated, it is clear that Onelli believed that cleanliness was good for zoo admission, and thus its coffers.

The zoo also contributed to the wider municipal focus on imparting hygiene to the city and its people to combat social problems. Beyond recreating healthily while at the zoo, visitors could internalize lessons about hygiene and apply them to their daily lives. Observing the installations and reading the guide, they could not help but note the reiterated statements on the importance of hygiene and perhaps pick up some pointers on how to maintain it, such as only drinking clean water and disinfecting surfaces that were frequently touched.[35] In terms of this broader project, the zoo held a unique advantage: visitors came from all socioeconomic classes, included Creole Argentines and immigrants, and covered the entire lifespan from infancy to old age. While schools, neighborhood programs, or medical clinics could address specific portions of the population, the zoo (and Parque Tres de Febrero in general) had been conceived as a meeting point for a large swath of society. The zoo's many programs and publications also had the ability to convey information about hygiene to different audiences: the more erudite *Revista del Jardín Zoológico de Buenos Aires* published technical guides such as Dr. Pablo M. Barlaro's *Tratado de higiene práctica elemental* (Treatise on elementary practical hygiene) for scientists and educators, zoo guides were distributed freely to visitors and also used in schools, and verbal instructions and programs (discussed later in this chapter) were accessible even to those who were illiterate.[36]

The zoo also drew on its popularity and good reputation to collaborate with campaigns to inoculate the population against common infectious diseases. Since the 1870s, *higienistas* such as Emilio Coni and José Penna had expressed concern over low rates of vaccination.[37] Although smallpox vaccination became obligatory in Buenos Aires in 1886 and in the rest of the country in 1904, in the early twentieth century many people remained unvaccinated or only partially vaccinated, seriously threatening public health.[38] Especially concerning, some parents were reluctant to vaccinate their children because of rumors that the smallpox vaccine would make them ill.[39] If children fell victim to smallpox, hygienists believed, the nation's future would be in jeopardy. Given the large number of annual visitors, the quantity of families who attended, and the diversity of visitors' socioeconomic status, the zoo was a felicitous place to address the population's lingering hesitations over what appeared to *higienistas* as an obvious choice to vaccinate.

As with disinfection and potable water, the widely distributed guide was pressed into service to promote vaccination. In what must have seemed a utopic statement at the time, the 1908–1909 Buenos Aires Zoo Guide boldly announced, "Smallpox will disappear when all inhabitants are vaccinated and boosted."[40] In the rest of the booklet, single sentence notes on page bottoms informed visitors of the consequences of smallpox, including heart damage and even death, noting that the disease killed an average of three *porteños* a day.[41] Other pages of the guide listed the locations and opening hours of vaccination clinics, allowing visitors to read about the zoo's condors while learning, for example, that "Vaccine Office 1 is located at Calle Olivarría Number 1561 operating from 9 to 11 a.m. and from 2 to 4 p.m. (Boca and Barracas Parishes)."[42] The guide also directly addressed common concerns, such as the quality of the vaccine (top), if pregnant and nursing women should be inoculated (they should), and whether it was safe to bathe after vaccination (it was).[43]

The writer(s) also relied on emotive language and scolding, reminding zoo visitors that smallpox was "the only disease that sufferers should be ashamed of, because it is also the only one that can be avoided by getting vaccinated" and calling it "one of the most disgusting illnesses."[44] This language reflects a broader cultural shift: whereas disease was once mysterious and uncontrollable, scientific advancement paved the way for a growing linkage between health and personal responsibility.[45] There were now concrete steps individuals could take to prevent and cure disease. Not taking those steps was a moral failing that threatened the health of the individual and society.

More than a decade later, the influential Liga Argentina Contra la Tuberculosis (Argentine Anti-tuberculosis League) advertised in the zoo guide, listing four clinics offering free treatment as well as the products they sought to have donated.[46] These dispensaries largely served the poor, combining medical care with anti-tuberculosis education.[47] Advertising in the popular guide contributed to the Liga's public awareness campaigns and connected zoogoers to available resources for fighting one of the most important causes of death in early twentieth-century Buenos Aires. Given that the guide was also sent to provincial schools across Argentina as educational material, it contributed to the spread of information about illness and vaccines in the interior.[48]

Onelli also worked to boost vaccination among his own staff, requesting doses from the city's vaccination program so Dr. Victor Widakowich could administer them directly to the unvaccinated while they were at work.[49] Many, if not most, of the zoo employees were poor, illiterate, and foreign-born, and thus exactly the subjects that hygienic interventions most targeted. Lack of time, information, or transportation were no longer excuses, and as such, Onelli hoped all the zoo's personnel would soon receive their certificates of vaccination and be spared the debilitating consequences (including missed days of work) of a highly preventable disease. From fresh air to variola vaccines, the zoo ideated and supported public health projects, collaborating in citywide efforts to protect the bodies, reproductive potential, and future productivity of all its inhabitants.

MENTAL HYGIENE AND MORALS

For *higienistas*, mental health and moral development were just as important as physical hygiene for reducing disorder. The zoo played a diversionary role: by attracting workers and their families, it would keep them away from other less savory spots. In 1905, administrator Alejandro Pavlovsky claimed that "In the Garden this public [referring to the working class] finds a pleasant walk and at the same time enjoyable and useful education, which undoubtedly influences their moral life, distancing many of them from the vices of the tavern and gambling."[50] As such, it was important to keep adding new animals and attractions "so that interest never wanes and [the zoo] continues to serve as a diversion for those who would otherwise go to the races, gamble, or go to drinking establishments."[51] Indeed, Holmberg originally prohibited the sale of alcoholic beverages at kiosks in the park, keeping vice off the grounds; the 1920–1921 guidebook reminded readers that the Confitería del Águila did not sell spirits.[52] Workers who went to the zoo avoided the physical consequences of alcohol and the moral decay of gambling, spending time away from the family, and more.

Zoo programs and landscapes were also designed to calm the nerves of city residents buffeted by the struggle for life. In 1904, Onelli had a gazebo installed in the center of the zoo grounds, and for most of his tenure local military and police bands gave concerts on Thursdays and Sundays.[53] These concerts were regularly announced in the press, including in the socialist paper *La Vanguardia*, often under the heading "Música popular" (Popular music) or "Música para el pueblo" (Music for the people). In addition to being another attraction to keep city residents off the streets and draw them to the zoo's hygienic grounds, Onelli believed that music was "a sedative for spirits that have been in tension all week at work and in the difficult struggle for life."[54] As their lungs took in fresh air and their bodies enjoyed light exercise, he hoped workers' minds could leave behind their struggles and seek solace in music. Freed from their stressors, they would be healthier and less likely to fight. The fact that the musicians belonged to police and military bands would further orient visitors toward ideas of obedience, order, and patriotism. Onelli's zoo concerts aimed to undo the damage of modern life by calming nerves and creating joy.

Indeed, in his 1919 novel *Nacha Regules*, Manuel Gálvez incorporated a visit to the zoo into the regenerative arc of his main characters. In the novel, Nacha is a prostitute who wants to "be good" and Fernando Monsalvat is a young man whose disillusion with the inequities of the modern world leads to a mental breakdown. In her efforts to leave her old life, Nacha helps Monsalvat recuperate. During this period, his convalescence includes excursions to "Palermo, the Zoo, and Lezama Park."[55] Just a few paragraphs later, Monsalvat is cured of his mental weakness and Nacha resolves to marry him and become an honest woman. Even in the fictional world of *Nacha Regules*, the zoo is a healthy, positive influence that pulls people out of the city's *mala vida*.

One of the best real-life examples of the success of the zoo's moralizing projects are the Carnaval celebrations held on the grounds in the late 1910s. In 1916, Onelli organized the first "children's parades" at the zoo

FIGURE 1.3. Young children celebrate Carnaval in morally hygienic style on the zoo grounds. Behind them is the Templo de Vesta, built as a modern lactation space for nursing mothers. "El carnaval de los niños," *Plus Ultra* 2, no. 11 (March 1917). Image courtesy of the Ibero-American Institute–Prussian Cultural Heritage Foundation.

(fig. 1.3), soliciting donations from entities such as the Jockey Club, the cookie company Bagley Ltda., and the department store Gath y Chaves. Over 160,000 people attended the first celebration and photos of the children's costumes and parties were printed in all the city's major magazines. The celebrations themselves stressed moderation and morality, as evidenced by certificates from the 1918 version of the event. Aptly for a zoo celebration, a monkey was named "Momo, King of Carnaval" and award certificates attested that he had distributed them from his cage in the zoo. Each was decorated with a photo of the monkey draped over a female sculpture by Dolores "Lola" Mora that had been installed on zoo grounds in 1910.[56] The words and image were surrounded by a fretwork-style frame listing the "Virtudes del niño": *morals*, *study*, *culture*, *determination*, *sincerity*, and *integrity*. Like "health" and "hygienic," these virtues were not neutral terms. The image of Lola Mora's sculpture and the design of the certificate ground "culture" in the Classical arts while also underscoring that the virtues were classed, raced, and Judeo-Christian. Additionally, "study" and "determination" point to the capitalist context, where personal effort and productivity measured an individual's value. These words physically circumscribed the more recreational aspects of the celebration, such as the costume contest, on the page, reiterating the lessons to be learned at the zoo as well as the imposition of healthy limits. Although Carnaval was usually seen as a time for letting loose, Onelli's zoo Carnaval pulled children and their parents away from unhygienic celebrations and replaced them with fresh air and "good values."

Local observers noticed and praised these efforts. While Carnaval celebrations generally presented the right conditions for disorder—high spirits, crowds, the anonymity of costumes—the zoo's hygienic nature, lack of alcohol, and careful administration took them in the opposite direction, creating an edifying and safe celebration for people of all ages. Consequently, in 1916, "the Management of the Establishment was very congratulated and feted, both by the press and by individuals, because it was a true exponent of culture without the slightest note of incidents, which are difficult to avoid in a huge crowd that wants to have fun."[57] In 1917, Onelli had advertisements for his Carnaval events hung in different barrios around the city, continuing his efforts to make the zoo a democratizing space that was open to all.[58] Eighty thousand people attended during the first three days. As in 1916, a *PBT* reporter noted that despite visitors from various social classes, there were no incidents at all, in strong contrast with "the shameful scenes" of other celebrations.[59] Among the disorder of the city, the zoo was a refuge for peaceful, hygienic, moral fun for all ages.

Why was the zoo so successful at maintaining order during celebrations known for their excesses? Onelli hypothesized that his victory derived from the fact that the zoo's parades were explicitly family events, and good parents wanted to protect their children from crime and immoral behavior. Like *higienistas* of his time, Onelli believed that the strong family unit was protective against disorder; by offering recreation for all ages, the zoo strengthened the nuclear family. Additionally, children having fun spread their joy to the adults around them, and thus even the nonparents in attendance might have been inspired to be on their best behavior.[60]

Education was the final piece of the zoo's hygienic project. At the turn of the century, intellectuals and politicians were firm in their belief that educating the masses in science would reduce social disorder.[61] In a speech to future teachers at the Escuela Normal in 1907, Argentine scientist and zoo collaborator Fernando Lahille insisted that natural history, specifically zoology, "educates and strengthens common sense." Through observation and comparison, a man engaged in the study of zoology would learn information that would improve his welfare and that of his society in general, while also becoming better at governing his life and succeeding in his labors.[62] As with fresh air and exercise, well-being was tied to productivity, while individual benefit was inseparable from the collective good.

The zoo was an ideal place for these lessons, as visitors were in close contact with animals and could learn about their characteristics, behaviors, homelands, and more. The field trips mentioned earlier in this chapter initiated future generations in these topics, with in-person visits to the zoo supported by newspaper articles, books, and the zoo guide, all read and commented in class beforehand. In front of animal cages, there was "a chance to gain knowledge, in geography as well as zoology, history, botany, etc."[63] After, students would produce drawings or compositions about the experience. An example published in the *Monitor de Educación Común* reveals that a student, Río Raúl Franchi, viewed the excursion as both fun and an opportunity to reflect on poetry, the comparisons between men and animals, and human intelligence and progress.[64] The visits may have also contributed to the material well-being of students, as one group of teenagers found the chicken cages so interesting that they began discussing buying eggs to establish home chicken coops.[65]

Because of the zoo's broad appeal, it could surpass public education's reach by also drawing in adults. Indeed, in his defense of Argentina published for the 1910 Centenary, Ángel Bonetti explicitly connected the zoo, education, and morality, writing,

> This institution [the zoo] seems to respond to the desire for superficial luxury, but brings incalculable benefits to the morale of the working-class masses. Natural history that is learned by sight remains engraved in people's minds and stimulates them to reflect and make comparisons between the animals of one continent and another. It awakens the desire to know so many other things that are related to the climate, customs and general ethnology of other peoples.[66]

By educating visitors about animals, the zoo would inspire them to think more critically, ask questions, improve their common sense, and find more effective and moral paths forward as individuals and as a nation.

These initiatives would find new areas for expansion in the twentieth century with the development of the Sociedad Luz and schools for working-class adults such as the Universidad Popular de La Boca. Founded in 1899 by Mauricio Klimann, Juan B. Justo, Antonino Piñero, and Ángel Mariano Giménez, the Sociedad Luz aimed to teach workers about the natural sciences, hygiene, and other topics, always in support of socialist ideology.[67] Onelli was a frequent collaborator, giving several talks about zoology at the Society beginning in 1906. In a parallel to children's field trips, he also welcomed adult members on guided visits to the zoo. For example, in October

1913, 110 people from the Sociedad Luz visited, making it the organization's fourth most popular excursion that year.[68] Other Sociedad Luz visits were photographed and published in *Caras y Caretas* and *Billiken*. Onelli also spoke at the Universidad Popular de La Boca, founded in 1917 by Tomás Le Breton, with whom he frequently collaborated.[69] From infancy to adulthood, Onelli and his supporters hoped natural history learning at the zoo would provoke intellectual development, even among those with less formal education, and usher in a new era of national stability and progress.

Finally, many believed the zoo could teach kindness and care, attitudes that once inculcated could dramatically decrease crime rates. In 1908, the zoo hosted the country's "primera fiesta del animal" (first animal celebration), modeled after the British Animal Sunday. During the event, Ignacio Albarracín, director of the Sociedad Argentina Protectora de Animales (Argentine Society for the Protection of Animals), argued that the celebration was a "civilizing virus," and that animal protection would "begin humanity's march toward more extensive and freer horizons than those we have been fighting against for nineteen centuries."[70] What mechanisms would drive this transformation? Albarracín quoted a US policeman who claimed that in twenty-five years of experience, he had found that "a lack of humanitarian education is the origin of all crimes."[71] Paradoxical as it may seem, for Albarracín it was precisely animals who were the best source of this humanitarian education. He believed that teaching students to be good to animals would make them nicer to one another; therefore, school time should be dedicated to teaching students to respect animals and treat them well. European examples showed that children who received this education did not commit crimes.[72] Thus, he hoped that the celebration of animals at the zoo and in Argentine classrooms would eventually lead to a more peaceful society.

In 1911, the zoo installed plaques donated by José Pérez Mendoza, the president of the Sociedad Protectora de Animales—Sarmiento (Society for the Protection of Animals—Sarmiento, the second such organization in Argentina; the first was founded by Domingo F. Sarmiento and led by Albarracín). In line with the society's mission, the metal signs read "Sea compasivo con los animales." During his comments at the installation ceremony, Onelli insisted that being kind to animals was essential to progress, and that those who dismissed the society's efforts as frivolous in the face of human social problems were wrong. The plight of animals and the plight of the poor could not be separated, for being nice to animals trained people to be nicer to children and poor people.[73] A 1923 article in *Billiken* about the Día del Animal demonstrates that the same belief continued to circulate fifteen years later. It detailed various ways children tormented animals, concluding that it was dangerous to be cruel, for cruelty hardened hearts and diminished happiness. The final sentences were brutally honest: "Those who are cruel usually end their lives in prison. Be kind to everyone, dear children, and you will receive the reward intended for those who are good."[74]

Finally, the youth groups sponsored by the magazine agreed: these Comités Billiken included protecting animals in their recommendations for charity and good behavior. One group's president, Virginia Faciotti, spoke at a Día del Animal event at the zoo attended by over six hundred children. In her speech, she passionately cried: "Love animals; take care of them; don't mistreat them! Repay with kindness the benefits that they give you."[75] Another Billiken group even took Clemente Onelli's name in its title, honoring him as a model of upright behavior. When he died in 1924, its members hosted a "tribute of remembrance and affection for the former director of the Zoological Garden" a few weeks later. The president, Miss Zulema de Agostino, and several other members spoke.[76] From Albarracín's speech to the Comités Billiken, a clear line was drawn between cruelty to animals and human criminality. Going to the zoo and attending events there were among the best ways to learn otherwise and interrupt this pipeline.

In the late nineteenth and early twentieth centuries, the Buenos Aires Zoo was a lynch pin in efforts to reduce disorder, combat social problems, and ensure progress in the rapidly changing city. Hygiene was in its founding documents, and both Eduardo L. Holmberg and Clemente Onelli incorporated cleanliness into the zoo's grounds, practices, publications, and programming. Its location in the Parque Tres de Febrero allowed the zoo to be a salubrious alternative to the cramped space of the factory and the *conventillos*, with grass and fresh air to strength the population's lungs and prepare them for work. Attention to the cleanliness of the water, animal enclosures, and grounds kept visitors safe and gave a practical demonstration of techniques that could be applied to visitors' own lives. Zoo publications touted these efforts, as well as provided information about critical public health initiatives. And because hygiene went beyond physical health, the zoo also kept *porteños* away from morally dangerous activities (gambling, drinking), and provided peace and music to help zoogoers escape the stress of their daily lives. Finally, the lessons in natural history and being kind to animals learned during visits, guided tours, and special events like the Día del Animal equipped zoogoers with the tools for building orderly communities and regulating their own behaviors, ultimately leading to the better treatment of their fellow humans and avoiding criminality.

These hygiene projects were public-facing and interfaced directly with the masses. In the next section, I look at the behind-the-scenes scientific work carried out by zoo administrators and medical professionals. These experiments tied animal health and human health together, creating networks across institutions as diverse as the Jardín Zoológico, Sociedad Rural, Hospital Nacional de Alienadas, and Instituto Bacteriológico. These experiments stressed similarities across species, especially between working-class humans, "anormales" (people with intellectual disabilities), "locos" (the insane), and the *pensionistas* of the Zoo. Notably, the benefits for different species were unequal, and human health, particularly at a community-wide level, came at the cost of nonhuman bodies.

THE ZOO AS MEDICAL INSTITUTION

The Buenos Aires Zoo contributed to the health of the nation through collaboration with medical institutions. At first, its animals benefited from many of the innovations in hygiene that protected human visitors. Zoo animals often lived in close quarters that echoed those of the human masses and suffered equally from tuberculosis and other contagious diseases.[77] But while the chimpanzee Cónsul III had expressed concern about human *porteños* living in closed housing, he would have no such criticisms of the zoo. Onelli took advantage of Buenos Aires's mild climate and his hygienic knowledge and focused on housing as many animals as possible in open-air enclosures. He also rigorously controlled the quality of feed and sanitary conditions. These efforts paid off. By 1907, Onelli was boasting that good hygiene, food, and medicine at the zoo had permitted animal reproduction, even of species notoriously difficult to breed in captivity.[78] Furthermore, Onelli reported that zoos worldwide had a 12 percent mortality rate, but in Buenos Aires in 1914 the rate was only 3.5 percent in mammals, 3 percent in birds, and 5 percent in reptiles. He claimed this number put Buenos Aires among the "healthiest" of the world's zoos.[79] Notably, his innovations ran contemporaneously or even a bit prior to similar changes in US zoos, reaffirming Hochadel's thesis that the Buenos Aires Zoo was not merely a passive recipient of scientific knowledge from the so-called center.[80]

Zoo science was still in its infancy, so Holmberg and Onelli relied heavily on human medicine to treat their *pensionistas*. In 1893, Holmberg decided his own medical skills would be just as appropriate to care for an elephant as a veterinarian's, given that no vets in the area had any experience with megafauna from other continents.[81] Onelli asked the director of the Comisión Consultiva and *higienista* José Penna, dermatologist Dr. Sommez, and

Dr. González Herrera to help him with a polar bear with eczema, ultimately treating the condition by painting the enclosure red and installing red glass to mitigate the force of the sun's rays.[82] Later, a mother elephant was given "caritaxina Gibson" to augment her milk production because it was proven to work in humans, while a monkey suffering from seizures was compared to the children of "alcoholic parents . . . syphilitics."[83]

But the transfer of hygienic and medical knowledge was not unidirectional, nor was it even. The first line of the first issue of the zoo's journal was "A Zoological Garden is a scientific institution," and both directors stressed that animal science should be useful to broader human society. For example, in April 1912, Onelli dedicated an article to Drs. Julio López Mañán and José Penna in which he insisted that science should escape the lab and "gather useful corollaries for life."[84] The article then listed practical examples from the zoo of how to fight disease-transmitting mosquitos and flies. Onelli's best solutions included cultivating populations of fish, bats, toads, and frogs, all of whom had proven to be voracious consumers of the dangerous pests.[85]

Furthermore, through close association with doctors and laboratories in the city, both directors hoped that their experiences with animals could be applied to sick humans, particularly those in public hospitals. In February 1893, Holmberg wrote of having cured osteomalacia (a bone-softening disease common in zoo animals) among the felines in his care. He initially tried the existing "*human* treatment for the illness," but it did not work. Instead, he prescribed raw milk mixed with burnt egg powder, beaten together and ingested once or twice a day. The great success of this experiment led him to try it on children and affirm, "How many stunted boys ought to become pumas for a bit in order be subjected to the Zoological Garden's system!"[86] The failure of a human medicine in animals thus led to an improved treatment for humans; a circle of knowledge, testing, and creation. Holmberg also collaborated frequently with Dr. Roberto Wernicke, an initiator of bacteriology and microbiology in Argentina, the director of the Sociedad Rural Argentina's laboratory, and close follower of Louis Pasteur. Wernicke performed numerous necropsies on zoo animals and took the knowledge he gained from them back to his human patients.[87]

Following Holmberg's example, Onelli continued to work with practitioners from many of the city's institutions. Medical students frequently visited the zoo with their professors to study living specimens.[88] The death of an animal, while sad, was also "a true party for scholars," as each part of the corpse would be studied by doctors and scientists, and the skeletons often ended up in the Museo Nacional.[89] Comparative pathology and comparative anatomy were the most popular sciences, allowing the zoo to serve "for the benefit of suffering humanity."[90] For example, Dr. Roberto Solé from the Clínica del Hospital de Niños completed a necropsy on a drill monkey with myelitis; Onelli claimed this experience allowed Solé to cure an eight-year-old girl who came to the clinic presenting similar symptoms.[91] In 1909, Dr. Horacio Piñero of the Universidad de Buenos Aires's medical school installed "an office that is a kind of auxiliary station for higher research in comparative pathology" at the zoo.[92] Other medics who used the zoo's animals to improve human health were Dr. Luis Agote (who used animal hearts to work on blood transfusions), Dr. Beatti (tumors), Dr. Raffo (cancer), Dr. Krauss and the Laboratorio Bacteriológico (blood), Dr. Widakowich (embryology), and Dr. Errauzquín (dentistry).[93] In each of these cases, animals—in life and in death—served the human populations who lived around them.

Two specific examples of Onelli's collaborations show the depth of these human-animal ties and their relation to efforts to solve the *cuestión social* through medical science. First, perhaps inspired by his relationship with José Penna, Onelli had long been interested in the use of microbiology to improve animal health. He requested a microscope for the zoo in January 1905.[94] With Dr. Ricardo Lynch of the San Luis Gonzaga Hospital in Buenos Aires, Onelli then pioneered the development of coprology, or the microscopic study of feces, to determine why animals were sick and how to treat them.[95] While Onelli focused on how microscopic examination of animal feces could give insight on the ideal diet for each species, Lynch took these observations and applied them to humans through comparative pathology. As the two scientists insisted during a presentation at the Congreso Científico Internacional Americano (International American Scientific Congress) in Buenos Aires in July 1910, alimentation was one of the most important new topics for individual and social hygiene, and they had made great advances in understanding it precisely because of Onelli's access to animals at the zoo.[96] Lynch also believed that coprology "can be a non-negligible assistant in criminal medicine investigations," thus contributing to lowering crime in the city.[97] In his survey of world zoos, French zoologist Gustave Loisel praised the two men's work in coprology as well as the studies of comparative anatomy, ethnology, and psychology published in the zoo's journal.[98]

While Onelli and Lynch focused on the health of the body, Onelli and the psychiatrist Christofredo Jakob of the Hospital Nacional de Alienadas focused on the health of the mind. For example, Onelli noted that chimps and idiots had the same posture, saying "I simply note the observation so that if there is any correlation, it can be developed by the famous psychiatrist Dr. Ch. Jakob."[99] Onelli also supplied Jakob with skulls and brains from the zoo's animals for comparative studies.[100] These experiments resulted in a jointly authored book, *Atlas del cerebro de los mamíferos de la República Argentina: Estudios anatómicos, histológicos y biológicos comparados, sobre la evolución de los hemisferios y de la corteza cerebral* (Atlas of the brain of the mammals of the Argentine Republic: Comparative anatomical, histological and biological studies on the evolution of the hemispheres and cerebral cortex), which was first presented to the Congreso Científico Internacional Americano in 1910 and published in 1913. The first part of the project compared the brains of animals in the region, drawing largely from the zoo's collections and Onelli's expertise. The second part, still in progress when the *Atlas* was published, was being carried out at Jakob's "neurobiological institute at the Hospital Nacional de Alienadas," a mental institution for women founded in 1854. Both pieces of the study responded to the idea that "for modern scientific research, the path to a profitable study of the human brain, that solar center of the organic world, lies through the series of animal brains."[101] By studying animals, scientists could better understand the human brain, pinpointing the biological and social origins of illness, explaining cultural and racial differences around the world, and developing tools to ensure public health and protect the future.

The Buenos Aires Zoo was also intertwined with the development of criminology and carceral institutions in the capital. During this time, "Crime was defined in terms of conduct that threatened the health of the social body."[102] This "medicalization" of crime, as Salvatore and Aguirre call it, was emblematized by José Ingenieros's founding of the Instituto de Criminología (Criminology Institute) in 1907 within the walls of the Penitenciaría Nacional (National Penitentiary).[103] The institute was meant to scientifically investigate the causes of criminality and possible cures.[104] The introduction to Jakob and Onelli's *Atlas*—a sort of manifesto for comparative human-animal studies—was also published in the Institute's *Archivos de psiquiatría y criminología aplicadas a las ciencias afines*, suggesting that Ingenieros believed that their research on zoo animals would contribute to improving the city's psychological and criminal issues. Jakob published several other articles in the journal, including one arguing that the empirical study of psychiatry and other biological sciences could help prevent and rectify social and legal problems. In it, he referenced Onelli's work with monkeys with apraxia.[105] The fact that zoo animals were most often compared to criminals, the ill, and the insane suggests just how little power those groups had to dictate their own care, as well as how they were dehumanized for being different.

Onelli also interacted with the work of noted Italian criminologist Cesare Lombroso. Their interactions appear to have begun in 1905, when Onelli used the zoo's journal to publish a critique of Lombroso's work. In the 1870s and 1880s, Lombroso had compared the skulls of delinquents and Aymará Indians, identifying a particular groove in the vermis (middle lobe of the cerebellum) in both groups, as well as in the skulls of monkeys and lemurs. According to Onelli, "Lombroso categorically and absolutely declares that this vermian pit in man is a sign of atavism because in one case he found this pit in an adult man who had his lips shaped like a 'hare's mouth' [cleft lip]."[106] In contrast, Onelli detailed his own study of seventy-six Indigenous skulls from his personal collection, as well as monkey and lemur skulls from the zoo. His observations of a larger number of specimens led him to the opposite conclusion: the two Indigenous skulls with "grooves" belonged to "peaceful men we knew when they were alive," while the two that belonged to known criminals had "a very pronounced mesovermian crest in the center of the occipital scale" instead of grooves.[107] Although Onelli set out to confirm Lombroso's theory that the groove represented lesser evolution in lower races and regressions in criminals, his observations showed that there was no significant pattern to the groove's appearance or disappearance, thus disproving Lombroso's racist hypothesis.[108]

The two men also exchanged materials and articles. In 1908, Onelli printed a letter in the zoo's journal from Lombroso thanking him for sending a Tehuelche skull to his lab in Italy. The criminologist also noted "We have found treasures in your *Revista del Jardín Zoológico*: I have found very precious data in it that will help me in new studies of criminals."[109] Unfortunately, Lombroso did not specify which articles had caught his attention, leaving us to guess what animal data he found useful for his phrenological studies. The following year, he wrote again, this time asking for permission to reprint Onelli's study about spots on vertebrate animals in the journal he edited at the University of Turin.[110] Onelli and Lombroso's communications were part of a larger network of Argentine collaborations with the Italian positivist school of criminology, including Lombroso's translations of books by Luis María Drago and Cornelio Moyano Gacitúa and Enrico Ferri's numerous conferences during invited visits to Argentina in 1908 and 1910.[111] One commentator even humorously suggested that Ferri had a negative influence on the zoo, joking that Ferri's speech on women's inferiority was the final straw for the baby elephant who, upon hearing of its content, "said a melancholic and disdainful goodbye to this planet and let herself die the best she could."[112]

The archive also holds glimpses of more troubling ties with the Penitenciaría Nacional. In 1922, Onelli received a request from the laboratory manager of the penitentiary's medical corps requesting one or two "pregnant guinea pigs." In the letter, the writer appealed to Onelli's "scientific spirit" as well as his morals, concluding "I thank you in advance for your generous contribution to the benefit of the prison population."[113] The writer does not specify what the guinea pigs were to be used for, but the creatures have historically been common subjects of medical experimentation. Given the shared "carceral logic" underscoring the zoo and human prisoners, these sorts of exchanges are hardly surprising.[114] Indeed, in the penitentiary the possibilities of comparative anatomy appear to grow, as both prisoner and zoo specimen experience the conditions of captivity.

The Buenos Aires Zoo also contributed to the Penitenciaría Nacional's efforts to improve prisoners' "moral hygiene." A cornerstone of the penitentiary movement across the region was the idea that work was redemptive, an "agent of moral therapy."[115] In Buenos Aires, the Penitenciaría Nacional boasted twenty-three workshops, powered by inmate labor.[116] In the 1920s, the zoo ordered the bread for its animals directly from the prison's bakery, buying up to fifty kilos a day.[117] Zoo administrators also inquired about using the prison's printing press for guidebooks and other publications.[118] The zoo thus helped create the demand necessary to sustain the moralizing work of prison labor. Prison labor fed zoo animals, while zoo animals' bodies supported the development of criminology as a science and underscored its suggested social, and racialized, reforms.

These examples, which are but a few of decades-long collaborations between the zoo and some of the most important doctors and hygienists in Buenos Aires, make clear that scientific study of human health was highly dependent on animals born and/or kept at the zoo. While the animals did benefit from some innovations in human hygiene and medicine, particularly during the institution's earliest years, humans benefited far more. The advantages for humans also came at a higher cost, as most came about after animals' deaths. While there is no evidence that the zoo bred or purchased animals specifically for experimentation's sake, nor that animals' lives were ever intentionally shortened or illnesses induced, it did stock city laboratories with nonhuman cadavers and animal parts to be used in the development of human medicine. Without the unnamed animals at the zoo whose hearts were donated to his laboratory, Dr. Luis Agote may never have perfected the indirect blood transfusion with sodium citrate that saved countless human lives in future years.[119] Although elided in studies of prisons, hygiene, or medical science in Buenos Aires, Holmberg, Onelli, and the zoo's creatures formed part of the scientific-carceral-hygienic network seeking to eliminate disorder and modernize the nation.

SATELLITE SITES

While the Buenos Aires Zoo was a productive site for generating hygiene, *La Nación* noted that going to Palermo "for working people and for many others is a sacrifice that can only be made on rare occasions."[120] If the goal was to reach as many city residents as possible, particularly those with fewer resources and a greater need for transformation, something else needed to be done. In 1907, Intendente (mayor) Marcelo T. Alvear ordered Onelli to take the zoo directly to the working classes living in the southern barrios, part of a broader program of developing that sector of the city.[121] As Onelli reports, Alvear wanted to "bring municipal action, the improvements of progress and civilization, to those working-class and poor neighborhoods that are the best producers of the commune's wealth," and where the dense population "lives only with work fever, without incentives to cultivate the spirit, without honest distractions that provide pleasant rest to so many children of hard work."[122] Alvear chose Parque de los Patricios, inaugurated five years prior on the site of former slaughterhouses, as the location for the satellite zoo. About seven kilometers south of the Palermo Zoo, Onelli compared the barrio to the Villette Quartier of Paris or Whitechapel in London, two neighborhoods known for working-class and immigrant populations, prostitution, and crime. He described the residents as "restless, uncultured elements . . . *apaches* [slang for delinquent, of French origin] of the worst type," truly the least disciplined of city residents, but he was equally convinced that the zoo could "indirectly modify the dominant character" in a short time, transforming inhabitants into peaceful, honest, hard workers.[123] Like many of the hygienists and criminologists he worked with, Onelli tied together poverty, crime, and immigration.

Once the location was decided, Alvear pushed Onelli to have a primitive version completed and inaugurated in just twenty-eight days, evidencing the importance of the project. Superimposing the zoo over former slaughterhouses and dumps required cleaning, installing pumps and plumbing, laying down new streets, and generally improving the hygiene of one of the city's least hygienic barrios. To meet this deadline, zoo employees worked rain or shine, on their days off, and on holidays.[124] Onelli even used zoo funds to buy them lunch so they would not waste valuable construction time going home to eat.[125] Replacing the old killing yards and trash heaps at Patricios with the zoo's live animals and a hygienic place of recreation was a clear signal of the city's progress.[126] Onelli and his employees made it happen.

The Zoológico del Sur, as it was often called, remained in its embryonic form for several years until it was rebuilt in 1913–1914 and formally inaugurated in October 1914. In accordance with Onelli's plans and personal aesthetic preference, the new infrastructure was to be modeled on Greek and Roman architecture, bringing Classical glory to "the least artistic part of the city, where the municipality is now cleaning and sanitizing and building important drainage works." The intendente agreed that these constructions could be "one more teaching tool, keeping in mind that a Zoological Garden involves the idea of recreation, as well as general culture."[127] The park was then constructed with replicas of Greek and Roman ruins: the Claudio aqueduct (fig. 1.4), the Erechtheion, Jupiter's Altar, the Temple of Vesta, and more. The only significant difference between the originals and their Argentine copies was that the Parque Patricios buildings housed animals and their feed: birds lived in the Erechtheion, tigers and lions were housed in the aqueduct. The result was so splendid that a North American theater company filmed a movie set in antiquity on the grounds.[128] By 1920, the Zoológico del Sur received 120,000 annual visitors for whom the architecture could provoke calm feelings that would counteract the negative effects of industrial life in the modern city.[129] In Parque Patricios, then, as in Palermo, the physical improvement and sanitizing of the space went hand in hand with a cultural project mobilized through art and architecture.

While many of the hygienic benefits of the Patricios and Palermo zoos derived from the exotic alternatives they presented to everyday life—green spaces, neoclassical architecture, and imported animals such as lions and giraffes—another line of zoo programming focused on the domestic food animals who lived side-by-side with many working-class Argentines. Milk was a frequent target of reformers in Buenos Aires, essential for children's development and far too easily made noxious by unhygienic practices. Throughout the early twentieth century, new controls were placed on wet nurses, and milk dispensaries were founded around the city.[130] To complement these initiatives, in 1911 Onelli wrote a memorandum to Intendente Anchorena in which he proposed developing model milking parlors at the zoo and across the city. These *tambos modelos* would architecturally resemble those of Switzerland and serve two intertwined purposes. First, all milk at the *tambo* would be rigorously and scientifically tested multiple times as it traveled from udder to cup. These analyses would ensure that quality was consistent and that no germs were present. Second, unlike other *tambos* already existing in the city, these would be open to the public and

FIGURE 1.4. At Parque Patricios, a building designed to look like a ruined Roman aqueduct housed animal cages. Clemente Onelli, "El Parque Zoológico del Sud," *Revista del Jardín Zoológico de Buenos Aires. Época II* 10, no. 39–40 (December 1914): 142. Image courtesy of the Academia Nacional de la Historia de la República Argentina.

would demonstrate the latest in hygienic, efficient milking.[131] The precautions taken at these *tambos* to ensure their asepsis included intense washing of hands, cups, and cow udders.[132] Visitors could then go home with a bottle of milk and a new toolkit of practices and knowledge to be applied to their own cows. Onelli connected his proposed *tambos modelos* to national goals, asserting to Intendente Anchorena that they would ensure "the health and robustness of a nucleus of children, a robust generation who, in turn, will be robust parents."[133] The insistence on children who would then have their own children demonstrates the long term, generative aspects of these projects.

A second, related proposal focused on introducing milking goat populations to the city. Proper technique and sterilization could eliminate bovine tuberculosis germs and impurities from cow's milk, but doing so eliminated some of the nutritive value. It was also difficult for city residents who themselves lived unhygienically to ensure the milk they got from their cows remained pure until consumed by their children. Faced with these limitations, Onelli recommended following Paris's example and developing municipal goat herds and milking parlors. Goat's milk was superior to cow's milk because goats do not get tuberculosis; children could be served milk

La inauguración de la Cabrería Municipal

Llegada del presidente el día de la inauguración, 17 de Octubre.

Edificio de la cabrería en el Zoo del Sud Parque de los Patricios.

El doctor Sáenz Peña y su esposa saboreando la leche de cabra

El doctor Sáenz Peña y la señora Lezica de Alvear.

El plantel de cabras destinadas á surtir á la cabrería.

FIGURE 1.5. The zoo-sponsored municipal goat-milking parlor would provide hygienic milk to residents of the Parque Patricios neighborhood, many of whom were poor immigrants. "La inauguración de la Cabrería Muncipal," *Caras y Caretas* 15, no. 734 (October 26, 1912). Image courtesy of University of Miami libraries.

directly from goats and it would ensure their "robustness and . . . health."[134] While some residents already drank goat milk, it was largely unregulated. To make his case, Onelli recounted watching a goat herder in Patricios scam poor immigrant women by pretending to milk a goat while pouring dirty milk from a bottle into their containers.[135] These swindlers endangered the health of Buenos Aires's most vulnerable. If the city were to invest in 2,000 goats, the zoo director argued, they would produce 1,500 liters of hygienic milk, enough to feed 3,000 children.[136] Onelli's arguments were persuasive, and the first cabrería municipal (municipal goat farm) was inaugurated at the Parque de los Patricios zoo on October 17, 1912, contributing to efforts to combat both tuberculosis and poverty (fig. 1.5).

Whether visiting the flagship zoo in Palermo or frequenting one of the model milking parlors, city residents were exposed to hygienic projects designed to improve their physical and mental health. However, as the examples in this chapter make clear, these projects did not have equal resonance for all groups. With its socially diverse attendance and aristocratic location, the Palermo Zoo featured general health precautions and morally fortifying music. More intensive interventions like *tambos modelos* and municipal goat farms were usually in working-class and/or immigrant neighborhoods such as Parque Patricios and La Boca. President Roque Sáenz and his wife attended the inauguration of the first municipal goat farm and drank milk directly from one of the herd, but they were not the principal targets of the project. Similarly, many of the zoo's medical collaborators worked at institutions for the incarcerated, poor, or insane, transitioning easily back and forth between studying captive animals and studying captive humans. In the following chapter, I tease out these social relations, examining how different social groups used the Buenos Aires Zoo and representations of its animals.

INTERLUDE

The hippos were eating their own shit again. It had happened before, delighting young visitors and horrifying their parents. Fritz and his wife would only defecate in the water of their pool. They would then eat it or throw it about, dirtying the water and the surrounding enclosure. After the first few such incidents, Onelli armed a zookeeper with a long skimmer to fish out the feces before its creators got to it. Unhappy with missing this special meal, the male hippo devised a way to subvert these human efforts. Now, as soon as the hippos got in the water "he would not remove his snout from his partner's tail until she had expelled an abundant helping."[1]

Feces was a huge problem at the zoo. The *pensionistas* produced enormous quantities of waste, requiring creative ways of disposing of it and putting the institution's hygienic projects at risk. Piles of animal feces offended the human senses of sight and smell and contradicted the image of the zoo as a *paseo higiénico*. Animal waste directly spread parasites and bacteria, both to other animals and to humans should it find a way into the zoo's always-struggling water supply. It was also a "fly farm," breeding insects that could transmit dozens of diseases, including tuberculosis, cholera, and typhoid fever.[2] The zoo thus presented a paradox: its animals were enlisted in the fight to improve hygiene around the city, but their embodied existence created conditions that, if not properly addressed, could significantly worsen public health. Maintaining the zoo's identity as a hygienic space that could counteract the filth and disease of the *conventillo* and the factory required constant laborious (and stinky) effort.

The sheer quantity of shit at the zoo was further complicated when animals expressed behaviors that diverged from human norms around

FIGURE I.1. The perpetrators. *Guía oficial ilustrada del Jardín Zoológico Municipal de Buenos Aires 1908–1909* (Buenos Aires: Gaibisso, 1908), 71.

bodily waste. Like hippos, tapirs ate their own feces when in the water.[3] Silky marmosets preferred to poop in their drinking troughs, while other small monkeys pooped everywhere and all the time.[4] Two species of zebras would dirty the fence between their enclosures with their waste, probably to mark their territory or for social bonding.[5] Anthropomorphs seemed to share the human desire to not dirty their own spaces, but in captivity that meant throwing their poop through the bars of the cages and into human paths.[6] For every cartoon or speech that used the zoo as a positive model, a real-life animal was engaging in behavior that directly counteracted best practices.

These somewhat disgusting observations reveal the cross-species tensions in hygienic projects at the zoo. As Onelli himself admitted, very little was known about why animals expelled waste the way they did.[7] To quote Japanese children's book author Tarō Gomi, "everyone poops," but at the beginning of the twentieth century the reasons for species variation were still a mystery.[8] Present-day science reveals that human ideas of hygiene, health, and disgust often contradict those of animals. Ingesting feces is dangerous for humans, but juvenile hippos eat their mothers' waste to stock their intestines, sterile at birth, with the bacteria they need to digest their food. A 2021 study on wild hippopotamuses in Africa also demonstrated that hippopotamuses' intense defecation loads their pools with hippo gut microbiota, creating a so-called mega gut. This "mega gut" might allow the sharing of microbiota between individual hippos, as well as affecting other members of the ecosystem.[9] In other words, living in a feces-rich environment seems to be an overall positive to hippo health, in direct contradiction with the human hygienic imperative to eliminate bodily waste as quickly as possible. What hippos need to survive goes against what humans need or consider sanitary. In other cases, like those of the orangutans, the general concept was shared but the captive contact zone of the zoo meant that when orangutans exerted their idea of hygiene, waste often ended up contaminating human spaces. Fashioning the zoo into a hygienic model meant counteracting these animal impulses, even at the cost of the specimens' health or happiness.

Humans themselves also challenged the zoo's hygienic projects. In 1905, Onelli wrote of a rural resident of Dolores who had come to the Buenos Aires Zoo specifically to obtain "a pound of camel dung to apply as a poultice to a tumor on his wife's chest." Onelli vehemently denied the request, but the man came back later that day, snuck into the camel enclosure, and stuffed his pockets with feces.[10] Humans also shared germs through bodily fluids, particularly saliva, and efforts to counteract the spread of illness through spitting, shared cups, and children's mouths on handrails were only partially successful. Zoo records are clear that visitors consistently broke the rules, even when admonished repeatedly. Other hygienic projects were thwarted by more pressing needs. In 1921, Onelli tried to make drinking water at the zoo more hygienic by supplying individual, disposable paper cups. But he had not anticipated the "cloud of children who frequented those popular walks and waited for that waste to take to their impoverished homes, where perhaps there would be one cup for the whole family; thus resulting in the fact that, in addition to the remains of ten or fifteen clients' saliva, they also brought home any germs that might be abundant on the ground."[11] Although his project was well-intentioned, it had overlooked the imperatives of poverty. Consequently, rather than understanding the early twentieth-century Buenos Aires Zoo as a utopic model of hygiene (as administrators hoped), we should view it as a microcosm of the struggles to improve public health happening across the city. While there was broad agreement that hygiene was the solution to most social problems, residents had their own needs and understandings of health and thus did not always use hygienic spaces and materials as they were intended.

Hygiene, or lack thereof, was also a potent weapon for the zoo's critics. Clemente Onelli's main antagonist throughout his directorship was Ignacio

Albarracín, president of the Sociedad Argentina Protectora de Animales. The two men were friendly but frequently butted heads over animal protection: while Onelli defended his zoo, scientific experimentation on animals, and eating meat, Albarracín was a vegetarian who once wanted to replace live zoo animals with taxidermy and argued that that the zoo's horses should wear little hats to protect them from the sun. As Urich summarizes, Onelli wanted animals to be treated well, while Albarracín advocated for "justice even for animals."[12]

In December 1910, Albarracín formally complained to the mayor that the zoo was sacrificing the equines destined to be big cat food in public view, traumatizing passing children and exposing visitors to waste and disease.[13] Onelli responded sharply, highlighting the hygienic measures his employees had taken: the sacrifice occurred in a well-ventilated building with a 3.5 meter wall where "the floor is always disinfected, the walls are waterproof, a marble table is always clean and makes the place look like an autopsy room, allowing the animals to be slaughtered without suffering and the cattle to be butchered in the most perfect hygienic state." The "remains" that could not be used were always carted away within one hour. If children happened to look through bars and see, he argued, then they were poorly behaved and should be reprimanded.[14]

Not content with this response, Albarracín gathered the signatures of "thirty-odd neighbors of the Zoological Garden, eyewitnesses and people of honor" attesting that "there [at the zoo] they attack public health and mistreat animals."[15] The complaint, which Onelli reprinted in the zoo's journal, vividly argued "that slaughterhouse is in the worst hygienic conditions. It is a true source of infection, the blood, urine, and manure go out into the street and gather on a clogged bridge on Acevedo Street, this bridge being a true microbial culture."[16] Onelli again defended himself and accused Albarracín of falsifying the neighbors' signatures, claiming some of the handwriting was identical. It is unclear if the neighbors' accusations were correct, a witch hunt, or an easy way to vent other frustrations with living beside a zoo, which to this day can be stinky and loud.[17] Onelli then went on the offensive, listing the addresses of those neighbors who he believed were themselves throwing wastewater into the public street. He concluded by reminding the municipality that the zoo provided drinking water to all who asked for it "for humanitarian and hygienic reasons," given that in its part of the city most of the drinking water came from illegal wells.[18] He thus turned the complaint on its head, weaponizing the allegations of poor hygiene against his accusers while simultaneously implicating the municipality in not ensuring clean water for all its residents.

FIGURE I.2. Zookeepers mockingly demonstrate what children would have had to do to see into the zoo's slaughterhouse. Clemente Onelli, "Cómo se pierde y cómo se hace perder el tiempo," *Revista del Jardín Zoológico de Buenos Aires. Época II* 7, no. 25 (April 1911): 85. Image courtesy of the Academia Nacional de la Historia de la República Argentina.

Despite seeming rather inconsequential, the fight was reported and discussed in many major newspapers, including *La Nación*, *La Prensa*, *La Argentina*, *La Gaceta de Buenos Aires*, and *El Diario*. They unanimously sided with the zoo. *El Diario* even published a photo of three zoo workers standing on each other's shoulders to peek into the slaughterhouse, mocking Albarracín's claim that the slaughter was happening in full view of the public (fig. I.2).[19] In the end, Onelli requested and received inspections by a commission from the city's other animal protection agency, the Sociedad Protectora de Animales—Sarmiento, as well as the municipal subsecretary Martín Urtubey. Intendente Joaquín Anchorena decreed on February 9, 1911, that Albarracín's claims were baseless, although he did recommend that Onelli put up a better fence.[20]

What does this squabble tell us about hygiene and the Buenos Aires Zoo? During the fight, *La Nación* accused Albarracín of inventing the hygiene issues because he wanted the zoo's lions to become vegetarians.[21] Although *La Nación* was clearly mocking him, there was likely a kernel of truth behind the claim. Albarracín, a fervent protector of animals, saw issues with the treatment of the zoo's specimens and those creatures sacrificed to feed them.

He was far from the first, as zoo critics have existed for about as long as zoos themselves. What Albarracín seems to have realized is that to grab the attention of the municipality and the public, animal protection would not be sufficient. How an old, lame horse died before being fed to a lion in a popular city attraction was not going to be a priority. But if that horse's death threatened public health, if it happened in a slaughterhouse that was "a true locus of infection" in an institution that improperly disposed of waste, transforming paths to nearby homes into "a true microbial culture," then city leaders would have to take it seriously. In this way, hygiene became the language with which to engage policymakers on a range of issues, as well as to mediate conflicts about much more than public health.

The zoo's founders and future directors conceived of themselves as forerunners in the fight for hygiene, administrators of a salubrious space that would impart knowledge, moral instruction, and physical benefits to the population. These advances would prepare the public to be good laborers and citizens and to raise even healthier future generations that would serve the nation. The truth, however, was that bringing this program to reality meant constant effort to overcome animal and human behavior.

CHAPTER 2

ARISTOCRATIC APES AND PROLETARIAN LLAMAS

Social Class and Labor at the Zoo

On November 11, 1909, a young woman named Jacoba celebrated her birthday in style. Her best friends were invited to her private residence for a five o'clock tea party in the British tradition. As was fashion, the afternoon ended with a dance. The society page of a local publication reported on the event, noting that Jacoba was "very spoiled by her friends."[1] The twist? Jacoba was an orangutan, her friends, howler monkeys and lemurs, and the paper the "Vida Zoo-cial" (Zoo-cial life) section of the *Revista del Jardín Zoológico de Buenos Aires*.

Throughout the early twentieth century, authors, visitors, and administrators represented the Buenos Aires Zoo as an aristocratic space filled with elegant animals living lives of luxury, despite having fur, fins, and scales. In a wide variety of texts that spanned genres and ideologies, writers consistently described animal enclosures as mansions and the zookeepers as servants who fulfilled the residents' every desire.[2] Some specimens, particularly anthropomorphic apes, were even dressed up and depicted engaging in upper-class behaviors. For example, the chimpanzee Cónsul, "from a noble family in Borneo," traveled to Buenos Aires in "a first-class cabin" and "was received with the honors appropriate to his hierarchy." The children's magazine *PBT* included a photograph of him seated at a table on an outdoor patio at the zoo, wearing a suit and hat and enjoying an aperitif. A bow-tied human waiter stood at attention, ready to serve.[3] Even less charismatic species were included in the animal elite: the zoo's royal chinchillas appeared in *Caras y Caretas* captioned "S.S.A.A.R.R. Las chinchillas" (Their Royal Highnesses. The chinchillas) (fig. 2.1).[4]

At odds with the homogeneously aristocratic animal population, human visitors to the zoo hailed from all ranks of *porteño* society. The Parque Tres de Febrero, home of the original zoo, was ideated as a democratic space that residents of every ethnicity and social class could enjoy. In his report to President Nicolás Avellaneda, park founder and former president Domingo F. Sarmiento highlighted that function by proclaiming, "Only in a vast, artistic, and accessible Park will the *pueblo* be a *pueblo*: only here will there be no foreigners or nationals, neither oligarchs nor plebeians."[5] Furthermore, as Juliana Adelman explains, zoo animals unite people from all social classes through their widespread appeal and by functioning as an oppositional pole: rich and poor humans have their differences, but form one block against zoo animals' animal-ness.[6]

Within a few years, the Buenos Aires Zoo attracted people of all social classes, but Sarmiento's desired erasure of differences failed to materialize. Elite families promenaded on weekdays while domestic servants and factory workers descended en masse on Sundays, provoking complaints from the wealthy. Writer and caricaturist Ramón Columba emphasized these class differences in a 1923 series of sketches (fig. 2.2). In one image, a well-dressed young woman happily partakes in the zoo's popular pony ride, while just below children without money to spare resign themselves to galloping on the iron steeds of park benches. Still other children in the drawings are at the zoo to work, such as the boys that supervise the juvenile cavalcade despite being barely taller than the ponies themselves.[7] As Columba's drawings make clear, nearly everyone could go to the zoo, but existing class differences persisted on its grounds.

In this chapter, I examine the interplay between the ubiquitous trope of the aristocratic zoo animal and the diverse and often conflictive relations between human social classes both at the zoo and in the city at large. In Buenos Aires, the early twentieth century was a period of intense renegotiating of social belonging and power, marked by a growing middle class and waves of labor agitation and repression. The zoo was no exception. From specimens who died of fright when striking protestors on Avenida Sarmiento set off fireworks, to metaphorical animal aristocrats, to relationships between administrators and their poorly paid workers, class relations and labor disputes permeated everyday zoo life. Overall, while the zoo offered some material for promoting workers' rights in early twentieth-century Buenos

SU ALTEZA REAL LA CHINCHILLA

Puna de Atacama, patria de las chinchillas.

Después de diez años las princesitas grises me han devuelto la visita y se han dignado hospedarse en mi casa. Vinieron oficialmente, presentadas por el ministro plenipotenciario de su país, pero sin aviso y tuve que preparar de manera somera el aposento de honor, mi biblioteca. Sus pequeñas Altezas Reales son de gustos sencillísimos: hubo por lo tanto que quitar las alfombras y dejar el parquet liso: pero en Buenos Aires exigen como mueble principal un hidrómetro el que fué colocado y, para evitar en algo la saturación húmeda que exhalan los lagos, dos jarrones fueron transformados en depósitos de cal viva, siempre renovada para que absorba la humedad del ambiente. Como princesas exóticas del oriente asiático si no viajan con cocinero de su tierra exigen por lo menos el manjar de su patria lejana, una gramínea dura; un pobre comestible de anacoretas. En ese aposento una alcoba; un cajón de pino con paja y arena y por dosel un tejido de alambre de donde cuelga el duro forrage traído semanalmente desde Catriló, en la Pampa.

Mesdames las princesitas duermen hasta muy entrado el día, tengo el honor de arreglarle yo mismo su departamento y me cabe el alto orgullo de haberlas visto en cama. Me reciben y platican conmigo amablemente á toda hora después de las 4 y hasta las 12 de la noche. No oigo su voz pero claramente veo que hablan por el movimiento de sus labios y de su bozo; pues mis princesitas calzan un bozo cerdoso de diez centímetros de largo: corren á mi alrededor, se arriman á mis botines, se paran verticalmente y con sus manecitas agarran de la mía algunas fibras de raíz de salsifí con que las obsequio. Mis princesitas son presidentas honorarias y efectivas de todas las sociedades de temperancia del mundo:

Las princesitas grises.

S. S. A. A. R. R. Las Chinchillas.

La alcoba de las chinchillas en Buenos Aires.

Puna de Atacama.—Un cazador de chinchillas.

FIGURE 2.1. Their Royal Highnesses, the chinchillas. Clemente Onelli, "Su alteza real la chinchilla," *Caras y Caretas* 15, no. 783 (October 19, 1912). Image courtesy of the University of Miami libraries.

FIGURE 2.2. Zoo visitors belonged to many different social classes. Ramón Columba, "Apuntes del Zoológico," *Páginas de Columba* 2, no. 7 (June 1923). Image courtesy of the Biblioteca Nacional Mariano Moreno.

Aires, its structure and its directors' political ideologies rendered it an essentially conservative institution only willing to work with labor in paternalistic ways.

CLASS CONFLICT AT THE ZOO

In the late nineteenth and early twentieth centuries, the acceleration of the Argentine economy, influx of immigrants, and numerous social changes led to a rapidly growing consciousness of social class, particularly as it pertained to labor conditions and rights. The middle class grew at an astonishing rate from 11.1 percent of the population in 1869 to 29.9 percent in 1914.[8] This growth, as well as the emergence of new political parties and labor activists, threatened the ruling oligarchy's power. The Partido Autonomista Nacional (PAN; National Autonomist Party), in control since 1874, saw challenges from (and was eventually defeated by) the Unión Cívica Radical (UCR; Radical Civic Union), which promoted a platform of universal male suffrage and was supported by a heterogeneous population dominated by that new middle class.[9] The Argentine Socialist Party was founded by Juan B. Justo and Nicolás Repetto in 1896; by 1904, Alfredo Palacios had become the first socialist national representative elected in the Americas.[10] The anarchist movement similarly prospered between 1890 and 1930, fed by immigrants from countries with strong labor traditions. The Federación Obrera Argentina (FOA; Federation of Argentine Workers), a collaboration between anarchists and socialists, was founded in 1901 and led the country's first general strike in 1902. A multitude of labor organizations and labor actions followed.[11] The ruling oligarchy struck back with violence, censorship, and laws such as the Ley de Residencia (1902) and Ley de Defensa Social (1910) that allowed the deportation of subversive foreigners and the refusal of entry to foreigners deemed criminals or anarchists, respectively. Tensions in Buenos Aires came to a terrible head in January 1919 during the Semana Trágica, a week of protests and repression that led to the death of hundreds of workers.[12]

In many ways, the Buenos Aires Zoo embodied these shifts and conflicts. The Parque Tres de Febrero, home of the original zoo, was built on the site of former governor Juan Manuel de Rosas's *quinta*. During his rule, Rosas had opened the space to elites, and it quickly became a popular upper-class promenade.[13] In the mid-1870s, Sarmiento's park and the zoo were intended to be "a remedy that had a positive impact on all possible aspects of urban life and that should reconcile the different social groups living in the city with each other as well as the city with the countryside," improving the lower classes and "appeasing social conflicts."[14] With the implementation of a low entry fee, lower- and middle-class Argentines now had access to this formerly prohibited space, a space that maintained a certain "aristocratic character."[15] The 1905 Law 4661, driven by socialist representative Alfredo Palacios and establishing Sunday as a day of rest, further facilitated mass access to the zoo as workers now had a full day for excursions.[16]

Contemporary accounts show that despite the founders' intentions, many upper-class Argentines were unhappy with the arrival of the masses at the Buenos Aires Zoo. For at least the first decade of the twentieth century, upper-class Argentines continued the Rosas-era promenading, going to the zoo to see and be seen. Bartolomé Mitre's conservative newspaper, *La Nación*, featured a semi-regular column in the "Social Notes" section called "In the Zoo." It reported on new zoo animals and special events, and often concluded with the names of the elegant people who had been seen strolling the grounds. The lists are a who's who of Buenos Aires's wealthiest and most influential families. On just one day in August 1905, members of the Varela, Mitre, Rodríguez Larrea, Bunge, Moreno, Aguirre, Pellegrini, Pereyra, Campos, Zorraquín Láinez, López, Pacheco, Lacroze, Achával, Huergo, Lanús, and Solé families were present.[17] Lower-class visitors challenged these activities and exclusivity, provoking numerous complaints from observers. British journalist and publisher John Alexander Hammerton admired the zoo's "large and interesting collection of wild animals" but did not recommend visiting due to the dust and the fact that "on Sundays it is so crowded with low-class Italians and the unwashed of all nations that one feels all the wild animals are not in cages."[18]

Hammerton was but one of many upper-class visitors who abandoned zoo grounds to the working class on Sundays and holidays.[19] The temporal division of the zoo was so pronounced that the poet Diego Novillo Quiroga reflected on it in his poem, "Día de Domingo" (Sunday day), part of the 1922 collection *Sabañones.* In the text, an aristocratic poetic voice bemoans the common people who invade public spaces on their day of rest:

> Oh, día de Domingo,
> díafano, transparente, luminoso.
> ¡Cuán simpatico fueras
> con tu cristal y añil, tu jaspe y oro;
> a no ser los horteras en las calles,
> los chicos en los biógrafos
> y guardas y domésticas en el Jardín Zoológico!
>   (Una cosa te salva en mi concepto,
>   oh, domingo: el Hipódromo . . .)
>
> Oh Sunday day,
> diaphanous, transparent, luminous.
> How lovely you'd be
> with your crystal and indigo, your jasper and gold;
> if it weren't for the philistines on the streets
> the kids in the movie theaters
> and the guards and maids in the Zoological Gardens!
>   (One thing saves you in my opinion,
>   oh, Sunday: the Hippodrome . . .)[20]

According to the poetic voice, the zoo, the movie theater, and even the streets were the exclusive play spaces of the wealthy on weekdays when the proletariat was ensconced in their factories and workshops; on Sundays, they were profaned. Novillo Quiroga was himself the son of rich landowners, but he intended his collection to be "malicious, daring, picaresque, sarcastic—caustic."[21] In a word, satirical. Rather than an endorsement of upper-class exclusivity, Novillo Quiroga invites the reader to ridicule those whose only concern for the poor was that they stay out of sight.

Novillo Quiroga's poem was tongue-in-cheek, but writers for *La Nación* did use its pages to critique lower-class members of the zoo community, be they employees or visitors, and to advocate for the status quo in the face of growing demands from workers. An article from June 25, 1914, chastised the zoo for continuing to employ a woman who "not only is disrespectful and inconsiderate with the public, but also boasts of it." The paper claimed that the employee, likely a women's toilet attendant, would demand tips and if refused, "becomes unbearable. With no attention to decorum, with no respect for anything divine or human, she utters insults that would make a saint indignant and curses that would make a cavalryman blush." The author wondered what influence the woman had over her bosses that they allowed her to continue working despite this behavior. He urged the zoo to put an end to it "for general culture and for the good name of the Zoological Garden, which, as one knows, is our children's recreation and the solace of many foreigners."[22] Unhappy with the attitude of a likely very poor woman, the paper publicly called for her firing because her behavior did not conform to the social norms of those who encountered her at the zoo.

Another author, signing as Tacuabé, used his experience at the zoo to invalidate the complaints of the growing Argentine working class. In his 1906 *La Nación* article, he describes people-watching from a bench at the zoo, taking in the diverse visitors of all classes, ages, and nationalities, a true microcosm of the city. He particularly notes the presence of "the small businessman, the laborer, the maid," claiming that "The *pueblo* is having fun, the *pueblo* is happy." Their presence in the aristocratic space of the zoo permits Tacuabé to construct a narrative of social mobility, especially for immigrants to Argentina. Although they were factory workers or domestic servants, they were able to afford the entrance fee and even the tramway fare to Palermo from other parts of the city. Their clothing was more than adequate and, he claimed, better than what they would have worn in their countries of origin. He then imagines following them home, where they surely have "abundant, hearty, and sometimes donated 'human' food," as well as savings and checking accounts, and maybe even ownership of a home or small business. These observations allow him to spin an elaborate fantasy of how through hard work, immigrants would continue to rise in social status while their happy, clean, educated children would grow up in the marvelous Argentine lands. Simply because the working class could visit the zoo, Tacuabé concludes that "in the Argentine capital, misery, frightful existences without food or a roof over one's head, are known only by the lazy and the idle, the impotent, the sterile, those who hope to obtain opulence by lecturing in tavern halls next to tables loaded with bottles and glasses." Honest workers had everything they needed, no reason to complain, and thus the "violent, hate-filled protests of the proletariat against capital" were unjustifiable.[23]

The socialist paper *La Vanguardia* quickly reacted, publishing an article by A. Regulo just days later that vehemently disagreed with the *La Nación* piece. Regulo argued that Tacuabé was disingenuous in going to the zoo to comment on labor issues: how could he assess the true condition of the proletariat on their one day of leisure and in an institution whose entry fee kept out the truly destitute? To understand the plight of the worker, he insisted, Tacuabé should have visited the workshops or factories. Because he based his observations on the zoo, he was clearly writing "with the obvious goal of making someone, I don't know who, believe that Argentina was an immigrant's paradise," an assertion Regulo found patently false. As such, Regulo dismissed Tacuabé's article as nothing more than oligarchical propaganda.[24]

Whereas some aristocratic Argentines lamented the "invasion" of Palermo by the masses and used the zoo to disparage workers, many others found it productive for advocating for moderate changes in social relations. These projects often depended on juxtaposing the aristocratic nature of the zoo with the human realities of the capitalist city. Many of the earliest world menageries were royal property and several of the most popular captive animals have long histories of being associated with royalty, including the "king of the jungle" (lion), king and emperor penguins, and, more specifically to Argentina, the "king of the Andes," *vultur gryphus*, the Andean condor. As mentioned, the physical location of the Buenos Aires Zoo and the enclosures that imitated the architectural styles of the upper classes in the regions of the animals' biogeography furthered the association. The bears lived like princes in the "Bear Palace" and the elephants lived in an elaborate Hindu temple. Commentators also created divisions between common, working animals and the more numerically rare and "unworking" zoo animals. As Clemente Onelli told journalist Augusto Vaccari, "Here [in the zoo], there are no donkeys, nor horses. . . . Why? Because outside the zoo they are plentiful . . . they are the majority, the animal plebs. . . . This place contains the zoological aristocracy."[25] Horses were ridden, donkeys pulled carts, and chickens laid eggs, but human visitors believed that most species in the Buenos Aires Zoo did not labor for or with humans. They were even free of the daily labors of wild animals, with no need to forage for food or face predators. Instead, they slept in enclosures built by humans and had zookeeper "servants" to bring them food and clean their waste. To the early twentieth-century Argentine public, a giraffe in a zoo enclosure appeared to have no responsibilities and could idle away the days basking in the sun, a perfect echo of the city's most wealthy inhabitants.

Commentators engaged this trope to comment on human social relations in the city. Despite Onelli's assertion that there were no animal plebs at the zoo, ponies, donkeys, and zebus pulled carts of food and excrement around the grounds, assisting the zookeepers who cared for the "more privileged" animals.[26] There was also a constant stream of children riding ponies, camels, and llamas. These attractions were so popular that *La Prensa* complained in 1917 that there were not enough animals to satisfy the huge number of children who wanted to ride, and even those with tickets were often turned away disappointed at the end of the day.[27] In writing and visual representations, these animals were explicitly associated with a politically conscious human proletariat at odds with the more aristocratic zoo animals. A 1900 cartoon in *Caras y Caretas* about animals' appreciation for art noted that as a group of human musicians walked past, the zoo's donkeys sung the "Marseillaise," the French anthem that became associated with revolution.[28] Four years later, a cartoon in *El Gladiador* depicted a congress entirely composed of zoo animals (fig. 2.3). The animal proletariat is represented by the horses who pulled the zoo's tramway; sitting in the "barra" (section for the public) they sing the "Himno de los Trabajadores" (also known as "The Internationale," an international workers' anthem) at the opening of the meeting and are called the "sacred voices of popular opinion." Their cries are silenced by the president, the aristocratic "Terrible Gnu." Other working animals at the session, including the llamas and zebus, question the traditional zoo animals' belief in the protective nature of the zoo and argue that for working animals, the zoo is a place of suffering.[29]

In these early representations, the working animals are associated with the proletariat and explicitly compared to their idle, aristocratic companions. Nearly two decades later, the image of the proletariat animal at the zoo endured, but they were now depicted as a proletariat even more oppressed than their human equivalents. In the previous twenty years, human laborers had fought for and won many rights, including the Sunday rest day. In contrast, animals worked harder than ever on Sundays, given that the human day of rest pushed droves of people to amusements like the zoo. As Goyo Cuello wryly noted, "children are lucky that they [the zoo's quadrupeds] have not invoked the Sunday rest [law]."[30] New legal protections for human workers were not matched by equivalent victories for animal labor, and in fact increased their (literal) burden. These representations ensured that readers saw the zoo animals not only as aristocrats, but also as part of

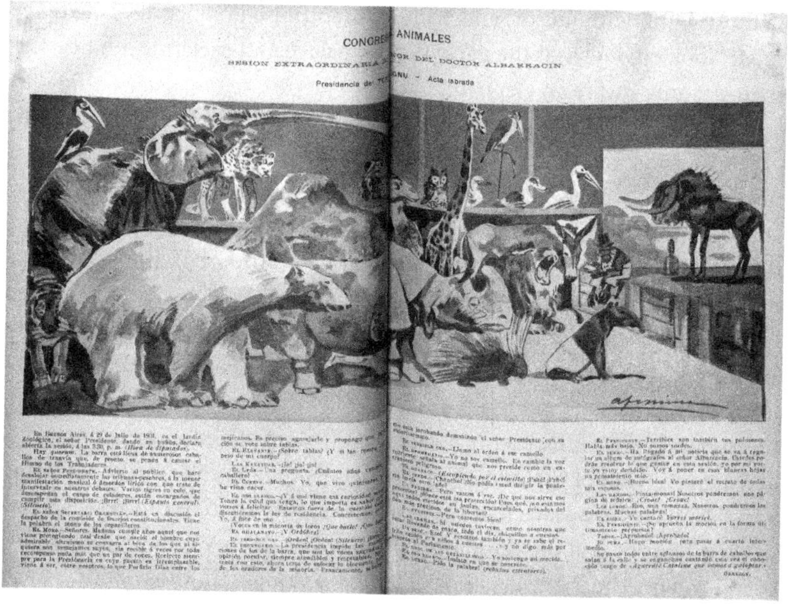

FIGURE 2.3. The foreign Terrible Gnu presides over the animal congress. Obregón, "Congreso de Animales: Sesión extraordinaria en honor del Doctor Albarracín," *El Gladiador* 3, no. 140 (August 5, 1904). Image courtesy of the Ibero-American Institute–Prussian Cultural Heritage Foundation.

a conflictive class structure that encompassed the same concerns contemporaneously bedeviling human society.

The gap between the aristocratic zoo animals' seemingly comfortable lives and the living conditions of the human residents of the capital was a common trope for social commentary. As discussed in Chapter 1, the working class often lived in crowded buildings that lacked hygienic measures such as ventilation and sanitation. After visiting the zoo and seeing the architecture, the Chilean traveler Rafael Sanhueza Lizardi asserted that the animals lived "better than many well-off people," while the Spanish writer Vicente Blasco Ibáñez alleged, "Many human beings would like to live in the mansions in which this Garden's beasts live."[31] Gabriel Carrasco's 1905 poem "En el Jardín Zoológico de Palermo" (In the Palermo Zoological Garden) expressed the same sentiment in verse: "Cúidanse tan bien aquí / Tan diversos animales / Que muchos hombres lamentan / No poder ser sus iguales!" ("So well cared for here / Are such diverse animals / That many men lament / Not being their equals!").[32] As an anonymous *PBT* author poignantly concluded, zoo visitors were always sad when they left, not just because the fun was over, but because they had to return to "their hovels without light, without joy, and without air." These poor visitors, he imagined, "look enviously at the pachyderms' palace and exclaim with heartfelt pain—if only I were an elephant!"[33]

This comparison was especially pressed into service during economic crises. In 1905, rapidly increasing rents marked the beginning of a period of unrest that would culminate in the 1907 Huelga de Inquilinos (renter's strike). In this context, Julio Castellanos satirically wrote that he was so desperate to find affordable housing that he considered asking Onelli to rent him the monkey cage. Unfortunately, his wife objected because, in her words, "there was nothing monkey-like about her."[34] Similarly, the effects of World War I and poor harvests meant that the cost of living in Buenos Aires jumped 65 percent from 1914 to 1918.[35] *Caras y Caretas* once again turned to the zoo to express concern. In a cartoon by Manuel Redondo, a man peers at a sign in front of the lion cage that details the big cats' weekly diet (fig. 2.4). Dismayed, he reads that the beasts are given 10 kilos of beef on Mondays, 11 kilos of veal on Tuesdays, 12 kilos of pork on Wednesdays, 13 kilos of mutton on Thursdays, 14 kilos of lamb on Fridays, 15 kilos of horse meat on Saturdays, poultry and fish on Sundays, and filtered water every day. The only logical conclusion? "Until the general crisis is resolved, I am going to request a pension in one of this hotel's cages."[36] The numbers were exaggerated (the animals fasted on Mondays) but the cartoonist used the zoo to effectively represent the tightening belts of the lower and middle classes.

While the "prensa burguesa" (bourgeois press), as the socialist paper *La Vanguardia* mockingly called many of its counterparts, instrumentalized the animals' standard of living to complain about prices that affected the middle class, *La Vanguardia* fully developed the juxtaposition of the aristocratic zoo animals and poor humans to argue for specific reforms that would benefit the proletariat. In 1906, the Italian-born journalist and photographer Esteban Dagnino described the Buenos Aires Zoo as an idyll where the residents lived in castles on the beautiful grounds. Like aristocrats, the animals "enjoy all the advantages of a privileged situation," including comfortable homes, the best food and water, and cleanliness. But rather than viewing these conditions as comic, adorable, or a source of Argentine pride, Dagnino argued that

> for those who are familiar with the infected caverns of our *conventillos*, a disgusting plague that cries out against the wealthy press's daily hyperbole about the so-called Argentine culture, for those who know how hundreds of proletariat families live crammed in there, there is nothing as useful and instructive as comparing that ugliness with the splendid neighborhood for beasts being built in Palermo. Reading one hundred volumes of sociology is less valuable than the reflections this comparison suggests.

FIGURE 2.4. The zoo animals eat better than humans. Manuel Redondo, "Noticias," *Caras y Caretas* 12, no. 801 (February 7, 1914). Image courtesy of University of California-Berkeley Libraries.

The fact that a different, supposedly inferior, species lived better than the proletariat truly illustrated the gap between rich and poor; those in power were more willing to spend money on warthogs than on workers. Dagnino conceded that the zoo had a place in Argentine culture, but "the care given to quadrumanes and cetaceans has the bitter taste of cruel satire." How many workers' houses could be built with the money spent on maintaining the zoo?[37]

On one occasion, the aristocratic animals "themselves" even took up the call to help the working class. In August 1922, *La Razón* published a letter from a lion that begged the municipality to invest in updating the zoo to be more like Hagenbeck's Tierpark in Germany, featuring fewer cages and more naturalistic landscapes.[38] A week later, a Bengal tiger from the zoo wrote to the paper to strongly disagree with the lion. He argued that if there was any extra money from the zoo it should go to "improving the conditions of many poor rational beings who, in terms of their housing, live worse than the animals." When families were stuffed into one room, promiscuity, illness, crime, and death proliferated, and "the spirit of rebellion and hate for Society" sprouted. As such, as long as humans were suffering, the animals should not "claim privileges, refinements, and other undoubtedly very pleasant things that we can do perfectly well without."[39] The kindly aristocratic tiger recognized his privilege and was willing to go without to help the humans living in misery in the capital and assist in avoiding class conflict.

Even more frightening to reformers was the comparative mortality of zoo animals and the Argentine poor. Buenos Aires's gentle climate and the care given to the animals, particularly the alimentary innovations that emerged from Onelli's coprological experiments, meant that the Buenos Aires Zoo achieved a low animal mortality rate in comparison with worldwide zoos. Administrators boasted of this statistic, but Dagnino used it to further condemn Argentine workers' circumstances. He wrote, "this statistic reminds me of the humiliating infant mortality numbers in Tucumán province.... If children die like flies in Tucumán, the zoo's beasts balance the scales with their very limited mortality."[40] Who could fail to be moved by the thought of children dying while hippos luxuriated? Indeed, Dagnino would likely have been unsurprised and unhappy to learn that in 1910, there was a lower annual mortality rate among the zoo's mammals than among its human employees (2.25 percent for mammals, 3.05 percent for employees).[41] As if anticipating objections, Dagnino concluded his article by noting that although the animals were captive and human workers were free, "what good does Freedom do if they [workers] must submit, under penalty of starvation, to whatever salary the master wants to pay them?"[42]

Finally, for socialist journalists, the concern expressed for zoo animals when they became ill or died also exposed how little the ruling elites cared for workers. When the orangutan Petronio and the juvenile elephant Phúa Victoria Porteña died in 1908, the papers published obituaries for them and Onelli thanked supporters for their sympathy. One year later, when Phúa Victoria's father Sayán became ill, the papers asked for prayers. On each occasion, *La Vanguardia* lambasted the exaggerated pity for the animals in contrast with the indifference to the same plights in the working class. Many worthy laborers died or lost limbs in the course of their work, but the bourgeois press did not eulogize them.[43] Even the explosion of a coal mine fire damp that killed hundreds of workers was barely noted while the death of a single elephant warranted dozens of pages of commentary. According to *La Vanguardia*'s journalists, this imbalance was underscored by the fact that both the upper class and the zoo animals were "paid for by the working people. The only difference is the price."[44] For them, the aristocratic zoo was yet more proof of capitalism's cruel inequalities.

Socialist reformers complained about the care and expense given to zoo animals while workers suffered, but over time they also came to collaborate extensively with the Buenos Aires Zoo. Like the more conservative *La Nación*, *La Vanguardia* reprinted articles from the *Revista del Jardín Zoológico de Buenos Aires* with some regularity, disseminating natural history knowledge. They also published announcements about concerts and special events on the zoo grounds and highlighted the inauguration of practical programs such as the model milking parlors and aviculture school. This programming, discussed in Chapter 1, aimed to improve workers' lives through hygiene and the imposition of middle-class understandings of behavior and leisure.

When Eduardo L. Holmberg was first removed from his position as director and replaced by Clemente Onelli, *La Vanguardia* dismissed the new zoo director as "a stupid sycophant."[45] In just a few years, however, the Socialist Party invited Onelli to participate in their educational institutions. The Sociedad Luz was founded 1899 by Mauricio Klimann, Juan B. Justo, Antonino Piñero, and Ángel Mariano Giménez to educate workers through a series of courses focused on "material, secular, objective, and universal knowledge."[46] Onelli spoke at the society at least twice and the group had his words printed as pamphlets that could be purchased by interested parties.[47] The Sociedad Luz also made frequent educational visits to the zoo with guided tours. During these visits, the zoo director would share humorous or interesting stories about the residents as well as information about useful animals and their commercialization. *La Vanguardia* would recap the tours for readers who had not joined the excursion.[48] Members could buy pictures from the outing for fifty cents in the Sociedad Luz's main office.[49]

Later, Onelli collaborated with at least two fledgling *universidades populares*. Like the Sociedad Luz, these universities aimed to educate workers, at least partially to avoid more radical forms of class protest.[50] Onelli was a founding member of the Universidad Popular de La Boca's board of directors in 1917.[51] In addition to guiding the institution from that position, he also gave classes about poultry husbandry, sharing the lessons learned from his experiments with the zoo's significant chicken collection and at the Practical Aviculture School he founded in collaboration with the municipality.[52] That same year, Onelli also spoke at the Teatro Colón for the Universidad Popular Bernardo de Irigoyen, later renamed the Universidad Popular de Mujeres.[53] Across these institutions, Onelli and the zoo served to educate workers in practical matters such as animal husbandry as well as providing a general knowledge and appreciation of natural history, particularly zoology. The hope was that these collaborations would improve workers' material conditions and thus health, lifestyle, and chances of successfully reproducing and being useful for the nation. They did not, however, encourage any radical reordering of society.

CLEMENTE ONELLI'S POLITICAL BELIEFS

No account of labor and the zoo would be complete without examining the director's beliefs as well as actual labor relations between administrators and employees. In terms of his own social position, director Clemente Onelli occupied a liminal space between the middle and upper class. Like one-sixth of the employed male population in Buenos Aires in 1914, he was an administrative employee.[54] Onelli was proud of his identity as an *empleado* and frequently self-labeled as such. He was even elected president of the Club Sindical de Empleados (Employees' Union Club) in 1917, giving a long speech that compared human society to a beehive. He argued that like worker bees, laborers should not reproduce, whereas his fellow *empleados*, like drone bees, had a double mission of "working for communal wealth and finding a way, with the modest personal remuneration they receive, to fulfill the noble mission of conservation of the species," that is, to reproduce.[55] They, the middle class, were the true base of society. At the same time, he married into the wealthy Panthou family and socialized with many of the oldest families in Argentina. He and his wife María Cecilia were listed in the social register.[56]

Publicly, Onelli refused to adopt any political affiliation. When speaking in front of socialists, he established a clear "me" and "you," finding common ground with his listeners but refusing to identify with them. More explicitly, in a 1916 speech at the Sociedad Luz, Onelli proclaimed "Every day I feel more socialist, more radical, more conservative, and at the same time more distant from all those sociological doctrines; therefore perhaps very different from you, exclusive socialists, forged on the anvil of need and in the crucible of abstract doctrines."[57] This ideological flexibility is probably why Onelli maintained his post as zoo director through PAN and UCR administrations, despite the directors before and after him being expelled due to political machinations.[58] Perhaps unsurprisingly then, Onelli's uses of animal metaphor and the aristocratic zoo are also fluid, if not balanced. In his writing, the aristocratic zoo and social classes for animals served to gently critique the upper class, allowed for some improvements for a deserving (working, cooperative) interspecies proletariat, and disparaged those that he viewed as threatening the fabric of the nation, especially communists.

After the turn of the century, middle-class magazines such as *Caras y Caretas* began to lightly criticize the upper class.[59] For the large number of middle-class visitors to the zoo, there had to be some delicious irony in imagining the upper class as animals, trapped in cages, and smelling of manure. The depictions of the bears living like princes and their royal highnesses the chinchillas surely provoked a frisson of satisfaction among the human population who knew themselves to be "poorer" but free. Onelli's "Vida Zoo-cial" (Zoo-cial life) column in the *Revista del Jardín Zoológico de Buenos Aires* similarly relied on aristocratic animals to lightly satirize the cosmopolitanism and stiff manners of the upper class. The column acted as the zoo's society pages, describing the lives of the zoo's *pensionistas* as if they were far above the hoi polloi. A party thrown by Australian black swans featured three juvenile females who "appeared very elegant under *gris-taupe* down jackets. Of a severe and stately elegance, the parents and relatives all attended the party in *velours noir.* The small ducks in their *toillette beige* were very admired for their *esprit* and elegance *tout a fait Parisienne.*"[60] The menu: purely French food. Other sections described the comings and goings of animals to their equivalent of the elegant beach resort of Mar del Plata.[61] In the columns, zoo social events are depicted as over the top and studded with foreign influences, an upper-class affectation that Onelli chastised in other articles.[62] If the insertion of apes and birds into narratives taken directly from the human society pages were not enough to indicate the mocking tone, the column is signed "Castigat Ridendo Mores," Latin for "one corrects customs by laughing at them."

However, Onelli much more frequently turned his focus toward the popular classes, drawing clear lines between a working proletariat that deserved protection and a radical subgroup that could only cause problems and disorder. His first public speech as zoo director was at the socialist Sociedad Luz in 1906. In it, he drew comparisons between working animals and working humans and advocated for a moderate socialism that would improve life for both groups. Departing from the assertion that animals were naturally destined to work for and be eaten by humans, he asserted "I maintain that we must apply humanitarian principles, the just theories of logic and a well-thought-out socialism, to these poor creatures, disinherited by fate and beaten by man, and try to make the slavery man has condemned them to more bearable."[63] Animals and the proletariat were natural workers, united in subordination but both deserving of fair conditions. Indeed, a boss's treatment of animals was a telling predictor of how he would treat his workers: "It is a proven fact that a man who mistreats animals, very frequently is also bad to his fellow man. . . . A foreman, a company manager who orders this horrible martyrdom for poor, defenseless beings like sheep, what consideration, what pity, what charity can he have for his workers?"[64] Through these reflections, Onelli established equivalents between human class interactions and interspecies relationships, drawing on socialist beliefs to advocate for kindness toward animals. The true socialist, he argued, should view his horses and other working animals as coworkers and defend their rights as he would those of his fellow human workers.[65] Whereas the socialist Dagnino argued in *La Vanguardia* that the good treatment of zoo animals only exposed capitalism's indifference to the worker, Onelli's formulation argues that "good" socialism would benefit all species alike.[66]

Despite this sympathy toward some animals and workers, Onelli believed there was another class so far beyond polite society as to barely merit consideration. Just as some elite visitors to the zoo complained of the dirty masses that swarmed the grounds, Onelli insisted that his aristocratic animals also distained the lowest of the low classes. In 1913, the zoo boasted of a gibbon named Cascarita who roamed the grounds freely. Onelli claimed she was always kind to children and well-dressed visitors, but frequently tried to bite the poor, be they visitors or employees. Like a human hygienist attuned to the worst of society's ills, the gibbon was particularly vicious with alcoholics. She once faced off with a zoo blacksmith in a scene "worthy of a poster against alcoholism." As Onelli described it, the worker "creeps shamefully among the plants in search of the poison that dominates him. Behold! The angel of temperance quickly descends from above: a black silhouette like a horrible gnome that stops him, shakes him, and, from the door's threshold, with the violence of the flaming guardians of Earthly Paradise, stops him and returns him to his work."[67] In this sense, the upper-class zoo animals were recruited into the fight against the rabble. Notably, the alcoholic that Cascarita targets is poor and, based on the zoo's labor records, likely an immigrant, validating hygienists' readings of Argentine social ills. The fact that even animals instinctively distained alcoholics appeared to prove the righteousness of those humans looking to eradicate such behaviors.

Onelli's depictions of the collections of small, common monkeys (not the more humanlike large monkeys and apes) further naturalized the existence of an immoral poor. Although he was one of the primary architects of the aristocratic zoo trope, he frequently depicted the monkeys as members of the least desirable sector of society and associated them with disorder and violence. Whereas in England, monkeys' promiscuity and violence was a metaphor for the uncolonized native, for Onelli, it stood for the urban poor.[68] He described the monkeys' common cages as *conventillos*, the crowded, multifamily tenements that hygienists viewed as sources of disease and disorder.[69] He argued that the similar material living conditions of small monkeys and the urban poor led to immoral behavior in both populations. He depicted the monkeys as shameless and wrote that of all the animals "that according to human criteria commit the sin of lust . . . Monkeys undoubtably take the cake."[70] The "tenement promiscuity" in which they lived meant that "good marriages are not possible. Everything is a hindrance, everything is a scandal, everything is gossip, everything is envy, and everyone's a farmer's dog."[71] They were also violent with each other, other zoo animals, their zookeepers, and the public, throwing feces and taking bites out of anyone that crossed their path.[72] These behaviors led Onelli to compare them to the lowest of the city's low, the nonworking, criminal element, and to use *lunfardo*—Italian-inflected slang—to describe them. One monkey demonstrated "the scoundrelly and violent brutality of the *apache* [slang for delinquent, of French origin] with his companion," while in general they behaved like *atorrantes* (Argentine slang for a bum).[73]

Through these representations, Onelli articulated two interspecies classes of poor *porteños*: those who worked hard and were malleable and those alcoholics and *atorrantes* who did not contribute to society. The former deserved care and protection; the latter were to be disparaged, just like Giorgi's lives to be protected and lives to be abandoned.[74] Onelli then mapped this gap onto the competing political programs looking to reform Argentine society at the time. In 1891, the Erfurt Programme of the German Social Democratic Party had called for the death of capitalism and a socialist takeover of the means of production. The 1889–1916 Socialist International followed suit, adopting this so-called maximum program as well as a minimum program focused on immediate demands, both economic and political. These reforms were not individually incompatible with capitalism but taken together would eventually lead to the proletariat amassing power and transforming society. In Argentina, the constituent congress of the Socialist Party elaborated a maximum program that argued that the working class was oppressed and forced into poverty by a few lazy capitalists; through political action, the proletariat would gain power and emancipate itself.[75] The congress also developed a minimum program, including an eight-hour workday, protective limits on women and children's work, gender equality in terms of salary, changes to the tax structure, universal suffrage, separation of church and state, et cetera.[76]

In his writing, speeches, and actions, Onelli juxtaposed "true socialism" (associated with the minimum program and docile, hardworking animals and people) and a malicious, maximum program of socialism that would unfairly benefit *atorrantes* like his monkeys or the alcoholic worker. Just socialism was premised on paternal protections and took root through actions like the programming Onelli himself provided at the zoo and socialist institutions in the city: It educated instead of protested, molding workers into a productive class while giving them just enough advancement to discourage more radical attempts at obtaining power. Again, animals provided a sheen of naturalness to his political positions: his observations, he argued, demonstrated that dogs "abhor the maximum program of socialism and allow only a minimum program."[77] Who could argue that man's best friend was wrong?

Birds provided the most fruitful examples of the wrongness of extreme socialism or communism, two terms that Onelli used interchangeably. In October 1907, he claimed that in nature, symbiosis was usually just parasitism, "the application of one of the points of socialism's maximum program, since the being that we will call capitalist kindly allows himself to be exploited by the other, a proletarian who does not produce or produces very little. The former and whoever observes them must firmly believe that both produce the same effort and that one cannot live without the help of the other."[78] The doves who built nests at the zoo and who "tend markedly to the comfortable symbiosis of the maximum socialist program" proved these assertions. When zookeepers fed the captive animals, the free birds would descend from their perches and gobble up the grain. Onelli claimed they had even managed to convince the camels and tapirs of "the truth of the old saying 'we're plowing'" and that symbiosis's mutual aid was fair: "we are helping you to eat and then we will fly freely in the air."[79] Returning to the image of zoo animals as an oligarchy fed by their servants, the free birds who do not "work" represent a lazy and corrupt proletariat exploiting the capitalists, in direct opposition to socialist premises.[80]

Onelli forcefully developed this image in "Gorriones sovietistas," originally published in the *Revista del Jardín Zoológico* and reprinted posthumously in the book *Aguafuertes del Zoológico* (1926) and in the illustrated magazine *Fray Mocho* (1928). The zoo director had a long-standing hatred of the sparrow, imported from Europe, that he blamed for disrupting autochthonous bird populations. In 1915 he wrote to the Intendencia Municipal advocating for sparrow hunting to protect local birds and crops. When a little girl wrote him in response and asked him to have mercy, he dismissed her concerns, saying the happiness sparrows brought her was outweighed by the tears they brought farmers.[81] In "Gorriones sovietistas," the birds, known for taking over other birds' nests, become "Soviets who have appropriated the bourgeois horneros' [ovenbirds, genus *Furnarius*] mansions." Bird behavior is interpreted as the conflict between the hardworking bourgeoise and the parasitical communist (itself a foreign invader and thus not Argentine) who "violates" what even children know to leave alone. Onelli concludes that the sparrow "is the communist of the winged kingdom, and he is so bitter and leathery he is not even good for cooking *polenta con pajarito* [an Italian dish]."[82] In these avian reenactments of labor relations, the communist is a usurper, a parasite, a freeloader, while the capitalist bourgeoisie works hard and is unfairly exploited. Although Onelli was happy to speak at socialist functions and to give guided tours to socialist groups, he apparently had no interest in the total transformation of society through revolutionary action. The reprinting of his work after his death ensured the lasting impact of his words.

One final example from the avian world demonstrated that not only was communism bad for the bourgeois, but it was also a tragic mistake for the proletariat. In "Idiosincrasias" from July 1909, Onelli explains how his *teros* (Southern lapwings) were content in their cage at the zoo until the day some wild lapwings flew overhead and told the captive ones about the beauty and abundance of the pampas.[83] Now, he complained,

> my lapwings can read the Marx of lapwings since they are no longer illiterate. They now perfectly know the lapwings' rights, they know that there are many others who are happier than them, and they have finally acquired mastery of their language. . . . Before they were poor dimwits who lived happily; and since the mortality of the lapwings' soul is certain for theologians

and materialists, I think that this life, which for theologians and materialists should have been their final paradise, has been made hell for them.[84]

In this case, exposure to the avian Marx did not improve the zoo animals' lives, but rather made them miserable. The analogy suggests that teaching the proletariat about class conflict and the promises of Marx would not set them free; instead, it would make them painfully aware of the bars of their metaphorical cages.

Onelli's suspicion of more radical labor groups appears to have been matched by those groups' avoidance of the zoo. Middle-class-led reformist socialist groups used the zoo grounds for recreation and education, but more extreme factions of the party stayed away. Furthermore, despite organizing picnics in numerous parks around the city, anarchists avoided congregating at the zoo. The middle-class and socialist press such as *Caras y Caretas* and *La Vanguardia* frequently mentioned the zoo, but the institution and its animals are largely absent from more radical papers. An examination of the anarchist paper *La Protesta* and its supplements reveals numerous images and stories about working animals, especially chickens, cows, and horses, and several pro-vegetarianism defenses of animals, but no significant depictions of the zoo.

The only explicit example I have found comes from the column "Música de actualidad," published in the anarchist magazine *Ideas y figuras* in 1909. The column purported to use black humor to cope with the misery and bad news of the present. In this edition, the pseudonymous author "Martín Fierro" says he spoke with Onelli in the zoo and learned about the opening of a new enclosure big enough to hold two elephants "*comme il faut*." Startled, the author responds, "¿Hay una jaula nueva en el Zoológico, / mi querido doctor, / y andan sueltos aún por esas calles / Figueroa y Falcón?" ("There's a new cage in the Zoo, / my dear doctor, / and yet loose in these streets / are still Figueroa and Falcón?")[85] Published on May 13, 1909, the article responded to Argentine Federal Police chief Ramón Falcón's crackdown on a May Day demonstration by the anarchist labor organization Federación Obrera Regional Argentina (FORA; Argentine Regional Workers' Federation), which resulted in the death of about a dozen workers. In the following days, known as the "Semana Roja," Falcón shut down socialist and anarchist papers and destroyed their printing presses, while also arresting numerous labor organizers. In response, FORA organized a general strike that only ended after intense negotiations with Figueroa Alcorta's office; Figueroa Alcorta himself had expressed support for Falcón.[86] Read in this context, "Martín Fierro" uses the image of the zoo cage to advocate for the imprisonment of Falcón and

FIGURE 2.5. A Labor Day fight between Clemente Onelli, Alejandro Pavlovsky, and the baby elephant. Cao, "La fiesta del trabajo en el Zoo," *Caras y Caretas* 9, no. 396 (May 5, 1906). Image courtesy of Cornell University Libraries.

Figueroa Alcorta, suggesting that if a cage big enough for elephants could be constructed, surely there was a place that could hold the chief of police and president. The zoo and its animals were malleable enough that they could be used by anarchists to support their goals, but the lack of articles and visits to the zoo hints that they did not find the institution particularly useful. It could be that the municipality-sponsored site was too connected to the government to be appealing or that, unlike other parks, the zoo had an entry fee. The zoo also encouraged the systematic and hierarchical ordering of nature, antithetical to anarchist principles. Or perhaps the aristocratic representation of zoo animals was powerful enough to keep the anarchists from viewing them as fellow workers and thus not useful as metaphor. Regardless of the cause, although both socialists and oligarchs utilized the zoo in support of their ideologies, anarchist factions did not follow suit.

This analysis of how different groups used the zoo and zoo stories suggests that overall, the institution served the interests of those looking to better the lives of Argentines through moderate reforms but did not lend itself to nor allow critiques that challenged entrenched power relations. A May 5, 1906, cartoon titled "Fiesta del trabajo en el Zoo" (Labor Day party in the zoo) depicted the limitations of zoo reforms (fig. 2.5). In it, a baby elephant (Phúa Victoria) pulls Onelli off the zoo administrator, Alejandro Pavlovsky, who holds a stick in a threatening manner. The caption reads, "the director and the zoo's administrator, amicably solving the *cuestión social*."[87] I have been unable to verify if there was a specific incident that inspired this cartoon but given the "Labor Day" title and allusion to the *cuestión social* in the caption, it seems likely the conflict stemmed from different interpretations of labor relations. Pavlovsky, a Russian who immigrated to Argentina in 1883, had published a study of the Russian nihilists earlier that year, giving platform to a group who encouraged an overthrow of the social order and distrust of authority, leading to a total transformation of society. The men's fight suggests Onelli vehemently disagreed. One month later Pavlovsky was replaced by Pedro Isla. Coincidence or consequence?

WHO WORKED AT THE ZOO?

A final facet of labor at the zoo is the composition and treatment of the institution's own labor force. This section is not intended to be a comprehensive labor history of the Buenos Aires Zoo, but it does begin to draw attention to topics identified by Alexandra Hui, Lissa Roberts, and Seth Rockman in the December 2023 issue of *Isis*, in which they argued for the importance of bringing labor history and history of science together.[88] Together with a discussion of women's labor in Chapter 4, this chapter also engages with Gabriela Soto Laveaga's call in that same issue of *Isis* to find precarious laborers in the history of science.[89] Who worked at the Buenos Aires Zoo? How was this labor raced, classed, and gendered? What were the power relations and how did they map onto contemporary understandings of human-animal interactions? Through recuperating this employment history, we find that Onelli's paternalistic, educative approach to the working-class visitors to the zoo overlapped significantly with his attitudes toward his own employees and, notably, the zoo animals themselves.

Using payroll records found in the administration's letter books from 1900 through 1924, as well as official correspondence, photos, and newspaper articles from the same period, we can piece together the labor force that kept the animals alive and the park ready for the public. First, all the employees who appear on the official payroll are men, with one exception.[90] Many zoo workers were immigrants. A 1917 list of the highest-ranking zoo employees shows that only nine of the sixteen were Argentine. That number includes Onelli himself, who nationalized sometime between arrival from Italy in 1888 and the publication of the list. The other seven were Italian, Uruguayan, Spanish, and French.[91] The payroll also reveals many Italian and Spanish names among the unskilled laborers. When Onelli tried to help his employees obtain a required police identification card in 1915, he ran into problems. As most of the employees were illiterate and foreign, their names were often so misspelled as to completely alter them. Other names were incorrect because the workers emigrated with other people's papers to avoid being conscripted into their home countries' militaries. Given that the men were generally "employees of very good behavior and who can only be defined as 'wretches,'" Onelli asked for the municipality's help in regularizing the situation and making sure they received the benefits to which they were entitled.[92] He repeated this plea in 1918, this time presenting a list of twenty-three employees whose names were recorded incorrectly.[93]

In terms of age, zoo employees covered the full life span. In letters to his supervisors at the municipality, Onelli repeatedly complained that his work force was very elderly, at times too old to be productive. At one point, he claimed that 40 percent of the laborers and keepers were "completely useless" due to their age. They had already been elderly when hired "because of the pious idea that the elderly must be protected by giving them light occupations with modest salaries to live on" and within a few years were in the unenviable position of being infirm but not yet in a financial position to retire.[94] A sad pattern emerges in the zoo's correspondence with the municipality: often, Onelli would pass on a worker's request for sick leave, following up only days later with a request for a replacement worker due to the same employee's death.

At the other end of the spectrum, the zoo was also dependent on child labor. Photos in the zoo guidebooks, the popular press, and in visual archives frequently show young boys, approximately eleven to fourteen years old, hard at work (fig. 2.6). Based on images, these children most frequently worked as Lilliputian train conductors, tramway operators, and pony and llama grooms, where they ensured the safety and enjoyment of their more fortunate peers. Many stayed at the zoo for years, causing budget issues when they came of age with "citizen's rights" and the zoo needed to increase their pay.[95] At least one unlucky child laborer died at the zoo. In 1911, a thirteen-year-old employee named Julio Signoretti perished when he was hit by the miniature train that circled the grounds. Signoretti had worked at the zoo since age eleven, caring for the llamas. The incident was written up in the crime magazine *Sherlock Holmes*, which sensationalized the tragedy: "The little locomotive that fulfills their [children's] satisfaction on their free afternoons by driving them around the grounds cut off a young life. The impression of this pain, amid the joys offered by Palermo's splendor, will late or never be erased from the minds of all those who witnessed the misfortune."[96] This tragic accident makes social divides at the zoo very clear. Many children visited the animals and enjoyed special attractions (which at one point also included a Ferris wheel), others had only enough money for admission,

CELEBRANDO EL AÑO NUEVO

El pe·sc~al del Jardín Zoológico

FIGURE 2.6. Zoo workers, including numerous children, celebrate the New Year. "Celebrando el año nuevo," *Caras y Caretas* 8, no. 827 (January 7, 1905).

and still others toiled to make these experiences possible for their peers. When magazines and zoo guides trumpeted the institution as a children's paradise, we must wonder how child laborers like Julio Signoretti and his colleagues experienced it.

Finally, for both children and adults, zoo work was poorly remunerated. The few men who worked in administration made decent wages and were granted tramway passes to make their way to and from work. The vast majority, however, made very little. At the end of 1917, 135 of the 158 workers on the zoo's payroll were making 90 pesos a month or less. Most earned between 65 and 80 pesos per month, far below the national average of $101.56 per month.[97] This included zookeepers, night watchmen, ticket takers, office assistants, mechanics, and laborers such as bricklayers and gardeners. A few even appear to have lived at the zoo, giving their address as the ticket office. Work at the zoo was often difficult and even dangerous. It involved cleaning feces, moving dangerous animals, staying up all night to monitor sick animals, and repairing heavy machinery. In reviewing these records, we gain a clear understanding of the precariousness of life for laborers at the zoo and the tensions between those who paid to enter the zoo and those paid by the zoo.

Onelli's relationships with his own employees largely echoed his theoretical position vis-à-vis labor as detailed in his speeches and writing. He is quick to dismiss employees who do not work hard or who show up drunk, but equally quick to defend those he views as "docile, obedient, and disciplined people."[98] He frequently served as an intermediary between his workers and the municipality, arguing for the reconciliation of names or benefits for the former child laborers who were now legally adults. He also helped resolve budget issues and advocated for higher wages, noting that the keepers had dangerous jobs and that the conservation of the zoo's important collections was directly dependent on their willingness to work. In October 1916, he wrote to the Intendencia to complain that his staff had not been paid on time, insisting that they pay out whatever money they could to the "neediest subordinate personnel" on the following Monday.[99] Onelli also helped the families of employees who died or became ill. In 1910, he requested financial help for the wife of a laborer who had died after a year of service.[100] When another laborer, Federico Gualtieri, became disabled and "irritable with the public as well as his peers and superiors," Onelli had to fire him. However, he wrote to the municipality asking that Gualtieri's son be named as his replacement, "thereby giving this Establishment a useful worker and not cutting off resources from a family of eleven with nine minors."[101] Onelli's request was satisfied, and Fernando Gualtieri, son, appears on the payroll later that year. Repeatedly, his support for his workers was contingent on his assessment of their obedience, willingness to follow directions, and work. They were not the *apaches* of the monkey house and the streets, but meek and obedient laborers that deserved to be rewarded (or at least paid for services rendered!). As Onelli's words made clear, he felt a humanitarian and practical impulse to treat workers well but absolutely did not support a broader overhaul of society.

Interestingly, zoo laborers seem largely to have been unable or unwilling to take direct action to improve their circumstances. In 1918, employees at the Córdoba Zoo went on strike.[102] Similar action did not happen in the *porteño* zoo at any widespread level. The only record of zoo workers striking that I have found is when four laborers, Leondro Donadio, Cayetano Marino, Donato Caleano, and Gabriel Reynols, asked to be reinstated after missing work "on the strike days" in May 1917. In this case, Onelli advocated on their behalf. He argued that they came to work those days for half an hour and then left, suggesting that they were only absent because of fear of retaliation from union leadership. He claimed their absence did not cause any additional costs, for they were not replaced.[103] Once again, in Onelli's opinion those who would strike voluntarily are depicted as "other" and even dangerous, while his "good" employees allowed him to paternalistically take control of their protection.

In many ways, Onelli's attitudes toward zoo labor as espoused in his writing and in his interactions with his employees echoed his attitude toward his zoo animals. He worked hard to make the captive animals' lives as healthy and kind as possible. He pioneered the field of coprology, reduced mortality rates, and repeatedly advocated for good treatment of all animals, including in his speech at the Sociedad Luz. He even regretfully referred to himself as the animals' jailer, while insisting that it was necessary. Just as he protected individual workers while railing against the "maximum program of socialism," he strove to improve zoo animals' lives and lamented their captivity, while never once considering granting them their freedom. In the next Interlude, I turn the trope of the aristocratic zoo on its head to ask if zoo animals can be laborers, and, more provocatively, if an institution built on an interspecies hierarchy can ever support radical labor rights.

INTERLUDE

Palaces, servants, unlimited meat, and warm baths: zoo animals appeared to have it all. But the zoo's internal documents reveal a different story. Anthropocentric understandings of comfort ensured that enclosures were not actually lovely for animals. While humans might be impressed by a replica Hindu temple, the elephants who lived in it had no use for the statues of gods and goddesses or "inscriptions copied from authentic documents" that adorned it.[1] They surely would have preferred tree stumps, rocks, and wet and dry wallowing sites for exercise and grooming. Many enclosures were not comfortable even by human standards, particularly in the early years. Most of the mammals slept on concrete floors due to a lack of straw bedding; in later years, firewood shortages similarly jeopardized their health.[2] Food was also in short supply, and administrators constantly wrote to suppliers and the municipality to beg for the delivery of fodder, oatmeal, or horses for the carnivores.[3] When fodder did arrive, it was often infested with insects, leaving the collections "exposed to illnesses and death."[4] Notably, these problems proliferated during the exact same period in which writers were using the zoo animals' privileges to highlight workers' poor living conditions. How much of a difference was there really between the lives of the captive animals and the poorest humans? The falseness of these elements of the aristocratic zoo trope points to the need to also question the assumption that zoo animals in their enclosures did not work, and to examine how this false analogy precluded other interpretive possibilities with more radical potential.

Theorists and activists have begun to argue that animal work goes beyond that of farm animals. It "not only includes animals' bodily strength in a narrow sense, but also involves multidimensional embodied active, relational, and interactive processes through which they engage their thoughts, emotions, abilities to understand, assess, and communicate, as well as their agency."[5] At the zoo, animals' primary task is "being themselves," which sounds effortless but requires performing lion-ness or monkey-ness as humans expect them to be, within restrictive and highly unnatural conditions. Humans want to hear a lion roar and look fierce, but not see how it poops, sleeps, or—horror!—has sex. Furthermore, like other animals in the tourist industry they are "front-line service workers," "expected to behave in certain ways to satisfy the needs and expectations of tourists, regardless of how tourists behave to and around the animals."[6] Enclosures are designed to privilege these human-determined interactions, with scheduled feeding times, toys or other activities, and architecture that encourages certain behaviors or limits the animals' ability to hide from the public.

In the early twentieth-century Buenos Aires Zoo, for example, enclosures were designed to allow humans to see the animals, privileging viewpoints over animal comfort. Some zoo animals were pressed into even more explicitly performative work, such as the chimpanzees Cascarita and Bertoldo, who rode bicycles around the grounds and led children's parades during Carnaval. They were stared at, surrounded by a noisy public, and had matches thrown at them, and yet the aristocratic zoo animal trope denied any possibility of them laboring at all.[7] It transformed captivity and the production of enjoyment for human beings into a valueless activity coded as simple existence or laziness, covering up the sacrifices of animals in captivity.

In a period of labor agitation in Buenos Aires, we can ask if and how zoo animals protested, advocated for themselves, and changed the material conditions of their existence. Violette Pouillard defines animal resistance as relational, meaning "any animal action which directly or indirectly challenges the institution's supremacy and which, as such, is rejected or countered on the grounds of its disruptive potential."[8] Jason Hribal gives examples of working animals who play dumb, ignore commands, take unauthorized breaks, complain, and otherwise interrupt the flow of work.[9] At the Buenos Aires Zoo, where animals' work was performing themselves, specimens frequently hid or simply did nothing, frustrating visitors' expectations. An early visitor to the zoo praised it as "one of the most pleasant recreation spaces in Buenos Aires" but expressed disappointment that the elephant "is not as gallant as he should be with visitors." Furthermore, the Galapagos tortoises hid among the grasses to the point that "if they exist, we believe it only because Dr. Holamberg [*sic*] has assured us it is true," while the tigers, lions, hyenas, and pumas were "a boring spectacle," lying about in their cages, and "do not even manage to instill fear nor give an idea of their fierceness." The bears refused to show their faces, and the camels spent "all day long lying in the sun." They refused to get up to be photographed for *Iris* magazine, even when offered cheese.[10] In later decades, a "bad humored" Jack the bear would not cooperate with the Carnaval celebrations and had to be replaced.[11] Highlighting the human expectation of animal labor, Juan Carlos Huergo complained in *La Nación* that many animals demonstrated "inconsideration to the public that spends money on entrance to the Garden" by not doing anything fun and showing "truly insolent laziness."[12] He implies that the public is entitled to certain animal behaviors in exchange for having paid the entry fee, despite the fact that the animals themselves did not directly receive the payment.

This refusal to labor likely reflects both animals' innate behaviors and the conditions of captivity: big cats do enjoy lying about in the sun, and living in a concrete and iron enclosure would surely lead to altered behaviors. In at least in one case, however, Onelli explicitly linked an animal's behavior to a refusal to face the crowds: an elephant seal, tormented by people that threw things at him, appeared to have learned the days of the week. On Sundays (the day of largest turnout and most working-class attendance), "he spends the day submerged between two waters, to make himself invisible. At the usual bucket-full-of-fish time, he does not come out to eat lunch even if the guard calls him." On the quieter weekdays, he emerged at the normal hour to eat.[13] In response to humans breaking the rules of good behavior, the elephant seal had learned to go on strike, refusing to show up to give the crowd what they wanted: a glimpse and a target. Whether intentional or not, these animals disrupted expectations, provoking the scorn of visitors and forcing zoo staff to figure out new ways to make them to act.

Other animals employed more, shall we say, anarchist strategies, by going on the attack. The archive is full of these stories. In 1911, the peon Roque Abrossi was cleaning the buffalo enclosure when one got angry and charged him, knocking him to the ground.[14] Another suffered a vicious raptor attack when the normally calm bird sunk its claws into the man, ripping his clothing, and the keeper "had to grab him tightly by the neck and nearly choke him in order to get rid of him."[15] A 1913 article on zoo animals' diets included a photograph of five arms and hands, all wounded or scarred, and captioned "Tasty mouthfuls" (fig. I.3).[16] The public was not exempt from these attacks, with at least one visitor being soaked by water from an elephant's trunk.[17]

Finally, in the spirit of labor organizing and solidarity, at least one zoo animal in Buenos Aires dedicated herself to freeing her co-workers. One of Jacoba the orangutang's favorite pastimes was releasing other zoo animals

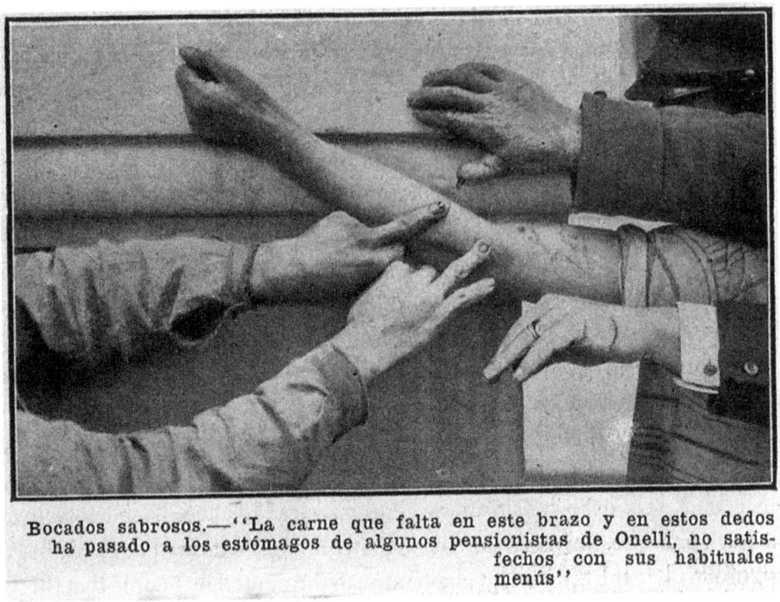

FIGURE I.3. Zookeepers show off the bites and scratches they have received from the animals in their care. "La comida de las fieras," *Fray Mocho* 2, no. 7 (October 3, 1913). Image courtesy of the Ibero-American Institute–Prussian Cultural Heritage Foundation.

from their cages. Upon her arrival in 1908, the zoo had to put locks on the titi monkey cages because she kept opening their cage doors so they could escape.[18] The longer Jacoba lived at the zoo, the more dedicated she became to helping her fellow primates. By October 1909, she was breaking cages to let the smaller monkeys out; her actions were so disruptive that Onelli had to move them to another area.[19] The following year, she freed a young mandrill and a group of gray-bellied night monkeys.[20] When the zoo installed stronger locks, she figured out how to open the door that separated the creatures' inside and outside enclosures, allowing them greater freedom of movement within the confined space. When that option was taken away from her, she took the top off a black howler monkey's cage, leading the zoo's humans to replace the nails with more difficult to remove screws.[21] "Organization and mutual aid are essential aspects in many animal cultures, including elephants, gorillas, and chimpanzees," according to Jason Hribal, but zoos limit those behaviors.[22] In her lock picking and Houdini-like escapes, Jacoba was returning to that culture.

Onelli praised his human workers for being docile and obedient, but his animal workers were often not. Their actions (or lack thereof) disappointed visitors and often required human employees to make material changes to their living environments or treatment. They were agents of social change. Nonetheless, these images of zoo animals as laborers are not present in the literature of the time, despite great awareness of labor issues by the public and the press. Why not? Were human exceptionalism and speciesism too powerful? Was the aristocratic trope—based on illusions—so compelling as to avoid it completely?

Furthermore, different social classes historically have had different relationships to animals, including the ability to exploit them.[23] How did the human proletariat and these animal workers interact and see each other? We have studies of modern zookeeper-animal relationships, but very little information about the early twentieth century.[24] It is possible that zookeepers, laborers, and the proletarian visitors did in fact feel some solidarity with the captive animals, but not the middle class that dominated the press and thus the record left for us today. It is certainly true that there were many parallels between the early twentieth-century Buenos Aires Zoo's human and nonhuman workers' lives, characterized by precariousness and oppressive labor conditions.

Do these parallels point to the possibility of interspecies labor solidarity? The archival record and the present are abundant in examples of kindly employees showing true care for animals, and some may have even assisted with an escape or two. Similarly, improvements for animals, such as when the Buenos Aires Zoo installed heaters in enclosures, also benefited the humans who spent time in or near their cages. Some animals have also acted in ways that alleviated keepers' burden, such as when an older "spinster" monkey sheltered and cared for the smaller titi monkeys around the clock.[25] Nonetheless, as the Argentine anarchists' reluctance to engage with the zoo suggests, the zoo's basis in captivity makes it difficult to imagine it as a space of worker liberation. Alasdair Cochrane defines "good work" for animals as "work which actively contributes to individuals' well-being and flourishing," based on pleasure, the exercise of animals' agency, and the recognition of animals as workers and community members.[26] A zoo depends on the forced labor of some species for the benefit of others, a relationship underscored by carceral fences and enclosures. We can never fully know zoo animals' will and degree of happiness, and thus cannot be assured that they undertake zoo labor voluntarily. Additionally, Jocelyn Porcher writes of the "shared suffering" of animals and human laborers in factory farms, articulating the mental toll of knowingly participating in animal suffering.[27] Jessica Sarah Rinland's film *Monólogo colectivo* shows that zoo employees also suffer due to their work with animals; even Onelli in the early twentieth century reported distress from his role as animal "jailer."[28] Five years after the death of the orangutan Petronio, he wrote that the ape's memory, particularly the fact that he seemed to waste away as a result of captivity, "still gnaws at my conscience."[29] Although humans and animals can labor together at the zoo and form lasting bonds, the hierarchical nature of the institution and its basis in captivity make it impossible to reconcile its existence with good work for all species, at least in its past and present forms.

CHAPTER 3

GUANACOS OR BENGAL TIGERS?

Zoo Collections, Acclimatization, and the Defense of the *Raza Latina*

In a 1914 comic strip, a Spanish immigrant named Sarrasqueta asks Mr. White, an immigrant from England, to help him pick a pet dog (fig. 3.1). Together, they board a tram to the Buenos Aires Zoo. Once in Palermo, they marvel at the animals' enclosures, including "the bears' great castle" with its hot, warm, and cold baths, and "the great elephant hotel, Indian pagoda style." In his broken Spanish, Mr. White exclaims, "How envious the workers will be! Here be much better than in the immigration hotel!" When the men meet up with Clemente Onelli, the zoo director asks Sarrasqueta what type of dog he would prefer: "How do you want it, greyhound, hound, terrier, retriever, pachon, bulldog, or fox terrier?" ("¿Cómo lo quiere, galgo, podenco, ratonero, perdiguero, pachón, bull-dog, o fosterrier . . . ?") In the punchline, Sarrasqueta mishears "fosterrier" for "forastero" (foreigner) and responds, "Yes, give me a foreigner as they're finer than those from this country" ("Sí, démelo forastero que son más finos que del país").[1]

From the immigrant characters and the mention of the immigration hotel to Sarrasqueta's preference for a foreigner, the comic, drawn by Manuel Redondo, engages with the realities of the country's rapidly changing demographics. In the second half of the nineteenth century, intellectuals such as Juan Bautista Alberdi and Domingo F. Sarmiento considered immigration

FIGURE 3.1. Sarrasqueta goes to the zoo to buy a dog. Manuel Redondo, "Sarrasqueta busca un perro," *Caras y Caretas* 12, no. 801 (February 7, 1914).

to be the solution to Argentina's problems. Immigrants (ideally from Northern Europe) would populate the pampas, be the labor force needed to spur economic growth, and bring the racial and cultural characteristics necessary to recreate European civilization in the River Plate. In sum, foreign bodies and foreign capital would usher in a modern future. By the time the Sarrasqueta comic was published, however, nearly half of Buenos Aires's population was foreign-born and this rapid demographic change with its attendant social and economic transformations left liberal elites disillusioned.[2] Many associated immigrants with crime, disorder, and disease, and advocated for laws such as the 1902 Ley de Residencia and 1910 Ley de Defensa Social which provided mechanisms for expelling or excluding those who did not support the nation.[3] Discomfort with the tangible effects of immigration in Buenos Aires melded with a growing cultural movement across Spanish America that rejected foreign influence and instead turned inward. Often linked in origin to the 1898 Spanish-American War and José Martí and José Enrique Rodó's literary work, this cultural nationalism reevaluated Spanish heritage, advocated for the spiritual over the material, promoted a Latin identity for the region, critiqued imperialism, and elevated the native and the popular (even if not concretely supporting Indigenous people or the working class).[4] In Argentina, it emerged in several interconnected fields, including education (Ricardo Rojas and Carlos Bunge), literature (Leopoldo Lugones, Manuel Gálvez), and policy (Manuel Ugarte, Vicente Quesada, and Roque Sáenz Peña).[5]

In this chapter, I analyze the Buenos Aires Zoo and representations of its animals in the context of concerns over immigration and imperialism and the concurrent rise of cultural nationalism. The zoo was a locus of encounter for the foreign and national, both human and animal. Many visitors were foreigners, from European luminaries (including Queen Isabel of Spain) to the working-class masses seen in Chapters 1 and 2, while others were Creoles of long-standing. Similarly, the creatures on display were native and exotic, with animals hailing from the banks of the Río de la Plata and the banks of the Ganges River. Commentators used the human categories of Creole and immigrant to characterize the specimens, a rhetorical move that allowed debates regarding species composition and acclimatization to be simultaneously scientific and social. At the same time, the design of zoo grounds modeled assimilation, patriotism, and a culturally based national identity to visitors. Overall, the Buenos Aires Zoo acknowledged that Argentina was a country of immigrants, but encouraged assimilation while defending the *criollo* to ensure that some fundamental national identity was not lost along the way. It thus fulfilled the conservative writer Manuel Gálvez's proclamation that "Modern Argentina, built based on immigration, that is, cosmopolitanism, can and must preserve a base of Argentineness."[6] The second director of the Buenos Aires Zoo, Clemente Onelli, also mobilized these representations as part of a pan–Latin Americanism designed to counteract US efforts to make cultural and economic inroads in the region. Contrary to the European framework that understood zoos as representations of imperial pretensions, in Argentina, the zoo and its animals were engaged as weapons for repudiating imperialism.

IMMIGRANT ANIMALS

Located in the urban capital of an expansive country, the Buenos Aires Zoo was dependent on animals brought from other places, be they the far-flung corners of national territory, sourced from European animal traders, or brought directly from Africa during the zoo's own expeditions. Zoo administrators, visitors, and writers anthropomorphized these animals and understood them in the context of two contemporary transitions shaping the nation: the consolidation of national control over the interior with its associated "pacification" of Indigenous groups, and mass immigration, primarily concentrated in littoral cities such as Buenos Aires and Rosario.

Imperial zoos staked their claims to far away colonies with exotic animals, but non-imperial nations such as those in Latin America did not have significant overseas territories.[7] Instead, the possession of native species marked the progress of internal colonization. In effect, records from the Buenos Aires Zoo's first two decades show that many specimens were native species donated by military men engaged in the exploration and pacification of the southern and northern frontiers. The very first report on the Parque Tres de Febrero highlighted that "frontier soldiers in contact with the desert, or port captains, and some provincial governors do not waste the opportunity to make it known that no matter how far they are from the Park, they understand its advantages, and wish to contribute to its beautification and perfection."[8] Julio A. Roca, leader of the so-called *Conquista del desierto* in Patagonia, was a frequent donor of animals between 1878 and 1904, including a condor, pumas, nutria, tapirs, flamingos, capybaras, and parrots.[9] Other donations came from military officials and explorers such as Lucio Mansilla, Francisco Moreno, and Manuel José Olascoaga. Other auxiliaries in these "civilizing" projects also supplied animals; a Franciscan missionary gifted a giant armadillo to the zoo just as the animal's Chaco homeland was being "pacified."[10] Native species at the zoo thus evidenced Buenos Aires's control over the interior.

Guanacos, armadillos, and rheas evolved in close contact with Indigenous tribes, but in zoo writing they are only rarely associated with the original human inhabitants of the region. Instead, they are repeatedly classified as *criollos*, a term used for people born in Argentina of European heritage. Indigenous people are thus erased from both Argentina's natural/national history and contemporary demography. This rhetorical move paralleled other attempts to invisibilize native peoples at the time, ranging from physical genocide to reclassifying Indigenous people as *campesinos*.[11] It also articulated an Argentineness common to many works of the Centenary Generation, opposing exoticism with a local identity that occasionally included traces of Indigenous people and cultures but largely skimmed over them in favor of a Spanish-associated identity.[12] Native species were thus neither Indigenous nor immigrant, instead representing a mythical (and frankly impossible) pure Argentine identity.

The parallels in the material conditions of circulation for exotic zoo animals and human immigrants also allowed for the institution's nondomestic *pensionistas* to be transformed into powerful vehicles for commenting on the immigrant experience, as well as their place in broader society. Argentine commentators repeatedly depicted non-native species of animals as immigrants with the attendant anxieties and opportunities. Of course, zoo animals did not choose to migrate; they were purchased and forcibly moved. A closer parallel might be the slave trade, but Argentine commentators minimized this relationship, likely because the erasure of Black Argentines and the history of the slave trade was quite advanced by this time.[13] In contrast, the discourse of immigration was ever present.

The story of Mimí the giraffe introduces us to some of those similarities. Mimí was purchased from a French trader in Saint Louis, Senegal, for 15,000 francs.[14] Argentine press depictions of her journey from her homeland to Buenos Aires explicitly paralleled her experience with those of human immigrants to Argentina (fig. 3.2). Like them, she traveled by ship to the port of Buenos Aires. Upon arrival, her "luggage" was inspected by customs officials.[15] She herself had to undergo health testing and quarantine, as human passengers had since the 1870s.[16] As a valuable specimen, however, Mimí was treated as a "first-class passenger" and allowed certain comforts denied to more common species, such as quarantining in her enclosure at the zoo instead of in the lazaretto.[17] As among humans, not all animal immigrants were equal.

Once in Buenos Aires, Onelli paraded Mimí through the streets of the capital as she made her way to the zoo, tracing the same radiating paths from the docks to the barrios created by human immigrants.[18] She

FIGURE 3.2. A giraffe named Mimí arrives in Buenos Aires from Senegal and is treated like a human immigrant. "La girafa 'Mimí,'" *Caras y Caretas* 15, no. 709 (May 4, 1912). Image courtesy of the University of Miami libraries.

then began a months-long process of adapting to her new environment, new foods, and new ways of existing that would have been surprisingly familiar to any human experiencing the culture shock of a new country. For example, Mimí's principal feed at the zoo was a mixture of oatmeal, ground corn, grass, bran, and salt, supplemented with hay, rye bread (sourced from the Penitenciaría Nacional), and a daily onion.[19] This was a far cry from a giraffe's native diet of acacia leaves, just as immigrants' kitchens and stomachs adapted to the lack of familiar products and introduction of new ones.

The similarities commentators noted between human and animal immigrants also included the emotional effects of migration. Mimí appeared to struggle to adjust to Argentina, and a *Caras y Caretas* writer reported general speculation that she was suffering from homesickness: "The newly arrived giraffe has not yet acclimatized to our Zoo's environment. They say she is sad, poor Mimí! No doubt, she has left some long-necked Rodolfo in Senegal, who pines for her, although without Puccini's music."

El animal no está enfermo,
según opina la gente:
¡es nostalgia lo que siente
la pobre Mimí en Palermo!
La consuelan los demás
animales del jardín,
diciendo, al ver su mohín:
¡Girafa, qué tris-tres-trás![20]

The animal is not ill
as people opine:
what she feels is nostalgia
poor Mimí in Palermo!
They console her
the other zoo animals,
saying, upon seeing her pout:
Giraffe, how sad you are!

Familiar narratives of human immigration, applied to animals, allowed *porteños* to try to bridge the species gap and provided frameworks for understanding the behavior of the exotic species that arrived at the zoo. Notably, many of these perceived similarities were highly anthropocentric and inappropriate for understanding giraffes or elephants, yet they persisted. For zoogoers and cultural commentators, Mimí embodied the immigrant experience despite being over ten feet tall and spotted. For the first four decades of the zoo's existence, this depiction of the exotic immigrant animal existed alongside that of the Creole—not Indigenous—native. In the next sections, I examine how these two animal representations were employed to grapple with very human questions about cosmopolitanism, immigration, and Argentine identity.

GUANACOS OR BENGAL TIGERS?

Early worldwide zoo administrators struggled to find the right balance between local fauna and exotic species.[21] What was the purpose of zoos?

Were they to delight and entertain? Were they to educate or for science? If meant to entertain, perhaps a zoo should focus on charismatic megafauna from far afield; an educational or nationalist approach might focus more strictly on local fauna. Practical and financial concerns also intervened: could an animal be acclimated to the zoo's climate? Were its preferred foods readily available? How much would transportation cost, and how likely was it that the animal would die in transit?

The Buenos Aires Zoo continually reflected on and interpreted the composition of its collections. On one hand, Argentine zoo and municipal officials spent a great deal of money on buying exotic species. A 1905 shipment from Carl Hagenbeck included tigers, polar bears, hyenas, zebras, camels, giraffes, and more, to the tune of 30,000 pesos.[22] These shipments continued semiregularly, and in 1921 the zoo sponsored its own expedition to Dakar, Senegal, to buy animals directly. These purchases were ideated and approved with the conviction that exotic species were necessary elements of a modern zoo and that unusual creatures such as rhinoceroses would attract larger crowds, thus padding their coffers.[23] The arrival of unique specimens was a major event, and magazines and newspapers around the country announced the new *pensionistas* with great fanfare.

Despite the scientific and economic reasons for bolstering the number of exotic species in the zoo's collections, the archive reveals equal interest in promoting domestic species in the collections and to the public. The municipal government ordered Holmberg and Onelli to increase their South American animal collections on several occasions in the late nineteenth and early twentieth centuries.[24] They complied. Despite some inconsistencies in identifying species and their biogeography over time, between 1904 and 1921 the percentage of South American species at the zoo grew from 34 percent to 61 percent of the total collection.[25] The actual population of native animals was likely much larger, as many of the autochthonous species were available in large numbers while exotic species like elephants or giraffes were more likely to be single specimens or pairs. The local press applauded these moves: *Caras y Caretas* reported, "another attitude of the Buenos Aires Zoo director, our distinguished collaborator, that must be applauded is his resolute inclination to provide American, or better yet, South American, or still better, Argentine animals to that modern Noah's Ark."[26] The article's author creates a hierarchy that assigns more value to species as their origins get closer and closer to home, contrasting with the demand for exotic symbols in imperial zoos as well as the Sarmiento-era preference in the Argentine capital for all things European. Foreigners such as Bishop Charles Warren Currier also positively commented on the extensive collections of Argentine animals.[27]

Why would the municipality and the zoo stress native species when the economic benefits seemed to lie with exotics? In practical terms, native species were easier to acquire and care for: they needed to be transported shorter distances, were more likely to be accustomed to a similar climate, and could be donated by travelers and residents in the territory. Those donations could foment residents' sense of connection to, and investment in, their city institution. Regina Horta Duarte has argued that in the early Bosque de Chapultepec Zoo in Mexico there was an oscillation between mimicking European zoos with exotic species and a "nationalist plan to proudly exhibit examples of the rich Mexican fauna," a dynamic also visible in Buenos Aires.[28] But given the Argentine social context and the frequency with which animals were represented as either criollos or immigrants, we can also read these scientific concerns as proxies for debates over cosmopolitanism. The concern with an excessive fealty to the foreign is plainly expressed in the Sarrasqueta cartoon presented at the beginning of this chapter, with the comment that foreigners were "finer than those from this country." In this long running comic strip, Sarrasqueta was a character often meant to be mocked. Here, his blind allegiance to the doctrine of foreign strength exposes him to ridicule. He has not seen the dogs on offer, indeed, does not even understand Onelli's options, yet immediately prefers the exotic for no concrete reason. Although immigration was already an omnipresent reality, the comic seems to say, there was no reason to disparage the local and overly praise the foreign.

The same hesitation toward an over-reliance on the foreign appears in commentary about the zoo's collections. As an early commentator wrote, the zoo "is poor in specimens of foreign fauna, and more unforgivably, it does not contain even one ten thousandth of our very rich fauna."[29] In one of the first issues of the *Revista del Jardín Zoológico*, from August 1893, director Eduardo Ladislao Holmberg similarly lamented the dearth of native species at the Palermo Zoo: "The Argentine Republic is very rich in species of Rodents, but the Zoological Garden does not possess but a small number of them because there is a vulgar and truly ill-founded belief that only rare animals should appear there, as if the Paca [*Cuniculus paca*], the Serelepe [Ingram's squirrel, *Sciurus ingrami*], the Tuco-tuco [genus *Ctenomys*] and many others were not, for most people, infinitely rarer than the maned Lion, the Elephant, or the Zebra."[30] Many Argentines, especially the city dwellers who visited the zoo, would have seen exotic species in books and circuses, but had next to no knowledge of species living closer to home. As such, Holmberg insisted that it was "a patriotic duty" to collect and display Argentine animals, teaching the public and giving new shine to animals previously derided as common or pests.[31] Similarly, Onelli insisted "if the Zoological Garden is an indicator of our culture and a place of spiritual education for the youth and for the masses, it should not just be a menagerie on a beautiful promenade where exotic species are piled up . . . but it should also be an exponent of local zoology, presenting examples of native fauna."[32]

Nationalist reformers agreed. In Leopoldo Lugones's educational manual *Didáctica*, which called for local and spiritual education, he called zoos and botanical gardens "the best classrooms," and lamented that more schools did not visit them.[33] Lugones himself also maintained correspondence with Clemente Onelli and published in the *Revista del Jardín Zoológico*. Carlos Bunge's nationalist primary school reader, *Nuestra patria: Libro de lectura para la educación nacional* (1910), included "Una visita al Jardín Zoológico" (A visit to the Zoological Gardens) among its patriotic texts. The book also included "El zorro sin cola" (The tailless fox), a story written by Clemente Onelli about a tailless fox at the Zoological Gardens.[34] Ricardo Rojas's plea for nationalist education as corrective to the heterogeneity of immigrant populations called for a focus on "the country's language, geography, morals, and modern history"; the connections between Lugones, Bunge, and Onelli suggest that the zoo and its native fauna were also among the educational elements that could unify that nation.[35] The zoo would consolidate national identity and inculcate patriotism by proudly displaying national fauna on equal footing with the visually more impressive exotic species.

Acclimatization projects at the zoo also examined the relative value of the foreign and the local through animal bodies, drawing similar conclusions as the debates regarding collection composition. Human explorers had long brought plants and animals home with them, or conversely tried to introduce species familiar to them into foreign territories. These efforts intensified and took on a specifically scientific sheen in the nineteenth century alongside European consolidation of colonial control over wide swaths of territory.[36] Known as acclimatization, these scientific efforts to intentionally adapt useful or interesting foreign species to the local habitat were systematized in specialized societies and on zoo grounds.[37] Acclimatization went beyond the adaptations of individual specimens to the new environment, focusing instead on introducing new species and trying to make them thrive despite (often intense) climate differences. The benefits were biological and financial. Acclimatization could "revivify extant faunas by introducing new species or by enhancing existing ones with exotic blood."[38] Accordingly, hybridization was a focus of many acclimatization societies.

Acclimatization of useful animals also brought economic rewards in the development of new food sources, textiles, and more.

The Buenos Aires Zoo engaged in numerous agricultural acclimatization projects in the late nineteenth and early twentieth centuries. Onelli was a *membre titulaire* of the Société Nationale d'Acclimatation de France and his library contained issues of the *Bulletin de la Société Nationale d'Acclimatation de France* as well as the *Bollettino della Società Zoologica Italiana*.[39] Over the course of his tenure, Onelli himself experimented with several potentially useful animals, including African ostriches, reindeer, alpacas, karakul sheep, and small mammals such as skunks and mink. His results were published in technical journals such as the *Anales de la Sociedad Rural Argentina* as well as the zoo's own journal. These Argentine experiences pose an interesting counterpart for existing histories of acclimatization, which have largely studied it as a tool of empire. For example, Helen Cowie has discussed alpaca acclimatization in England as part of Britain's informal empire in South America, arguing that it highlights the "complex networks of exchange and expertise at work within and beyond the British Empire."[40] Harriet Ritvo has examined acclimatization in English-speaking (former) colonies such as the United States, Australia, and New Zealand, while Baratay, Hardouin-Fugier, and Livingston write of acclimatization across Europe. As Pascal Grosse writes for the German context, "the acclimatization question became a self-reflexive code for the racial and cultural identification of white Europeans as whites in colonial societies."[41] A useful counterpoint, Pete Minard's study of Australia flips this paradigm, stressing the agency of colonial actors in driving acclimatization movements.[42]

None of these frameworks can explain Onelli's actions at the Buenos Aires Zoo, however, as he carried out his experiments largely after the decline of European acclimatization societies and outside the geographic sphere of empire. Instead, Argentine acclimatization discourse reflected the intertwined preoccupations of immigration and colonization of the pampas, particularly through semiotic transfer from exotic animal to human immigrant. In 1905, Onelli received four pairs of reindeer from Norway when Captain Carl Anton Larsen docked in Buenos Aires on his way to Antarctica. The circumpolar animals stayed at the Buenos Aires Zoo, relying on blocks of ice to resist the summer heat, from December 24 until January 13, when five set sail to Ushuaia, Tierra del Fuego, to continue the acclimatization experiment. Onelli's report on the reindeer in the *Anales de la Sociedad Rural Argentina* includes information about their diet, illnesses, and behaviors, as well as a photograph of the animals in their new home with the men who cared for them. He concludes that the acclimatization of reindeer was

Ganaderia exótica. Los renos en Tierra del Fuego

FIGURE 3.3. Reindeer in Tierra del Fuego. Clemente Onelli, "Ganadería exotica," *Anales de la Sociedad Rural Argentina* 40 (May–June 1906): 66.

successful and the surviving animals "will be the founders of a new Argentine livestock dynasty that will ensure food and therefore progress in certain regions with a difficult climate, poor vegetation, but rich in mines that await the work of man."[43] These immigrants were adaptable and economically profitable and thus would spur national progress. Furthermore, they would be specifically "Argentine." Notably, these processes would occur in lands that were currently little populated and not productive, meaning that they would not compete with existing industries.

Another welcomed immigrant animal was the karakul sheep, originally of Central Asia. In April 1912, a fur trader named Augusto Huber published an article in the *Revista del Jardín Zoológico de Buenos Aires* extolling the virtues of the breed and arguing for acclimatization efforts. He argued that "there is no doubt that the breeding of these animals will become a source of wealth for certain regions of our country that until now have been considered little less than useless due to their altitude and unfavorable soil conditions."[44] Like with reindeer, profitability drove the interest in facilitating acclimatization. Rather than import countless purebred karakul sheep, Huber recommended crossing the foreign breed with the native *oveja pampa*, which he argued was more valuable than commonly appreciated.[45] The karakul sheep would reinvigorate the local breed and could hopefully adapt to living in "the most scorned, or better said, the least populated regions" of the nation, including Río Negro, Chubut, and Santa Cruz.[46]

Onelli followed Huber's advice, and in July of the same year, he reported success breeding a dark, curly haired sheep similar in appearance to the karakul from an *oveja pampa*.[47] He then crossed those domestic sheep with karakul sheep donated by the emperor of Austria and Hungary, resulting in a creature that "although not achieving the perfection and size of the purebred ones, born in Tandil, were good enough specimens."[48] Through the processes of acclimatization and hybridization, Onelli had created a new species that would be economically valuable and able to fill the spaces of the underpopulated south, cementing national control over those regions and transforming barren lands into a driver of progress.

Although ostensibly about animal husbandry, these projects and the reports of their successes lent themselves to a broader secondary interpretation. Onelli and his colleagues drew parallels between animal acclimatization and the adaptability of twentieth-century European immigrants to a national context and debated the value of encouraging these adaptations through discussions of zoological acclimatization. In addition to the clear parallels between human and animal immigrants circulating at the time, the rationale for acclimatization—invigorating local species, generating economic benefits, populating remote territories such as Chubut, Santa Cruz, and Tierra del Fuego—mirrored the arguments for human immigration. By 1914, over half the capital's population was foreign-born, while northern and frontier provinces like Corrientes, Jujuy, Formosa, and Neuquén had less than 7 percent immigrant populations.[49] Reindeer and karakul sheep could provide the financial incentive for people to move to the provinces, while also themselves modeling the ability of foreigners to thrive in the interior.

Many Argentine articles about animal acclimatization at this time mentioned human immigrants, demonstrating that the social implications of these experiments were always present. For instance, in 1907, Onelli compared the adaptation processes of coyotes and jackals (North American and African/Eurasian canids, respectively) to how "the sober and poor immigrant, accustomed in his land to nourishing himself only with the bread whose daily and fundamental provision he entrusts to God in the Lord's Prayer, finds that this bread becomes meat in the American countryside."[50] Similarly, Onelli's close collaborator Cristofredo Jakob's *Tratado de biología general y especial para el uso de la enseñanza elemental, secundaria y superior en la República Argentina* paralleled the biological processes of animals and "human eugenics."[51] In the section published in December 1916, Jakob first insists that the evolutionary processes identified in animals post-Darwin could be extended to humanity: "The current European type, the Yankee, the gaucho, etc., are they not formations that history shows in their origin, transformation, and evolution?"[52] He later notes that migration is a behavior common to animals and humans and identifies it as a principal driver of this evolution.[53] These comparisons laid the groundwork for the argument that human immigrants could adapt, and, like the animals described above, become useful to national progress. Importantly, both Onelli's and Jakob's approaches to immigration inscribed it into the context of a highly controlled, scientific process whose primary goal was molding the body of the animal to fit local conditions and the biological, racial, or economic needs of the hand in charge.

Finally, as in debates over its collections, reports on the zoo's acclimatization experiments reiterated that while some animal and human immigrants were beneficial, they should not be encouraged at the expense of the Creole native. Echoing the argument of the Sarrasqueta cartoon above, collaborator and Museo de La Plata scientist Robert Lehmann-Nitsche declared in a 1907 article on the acclimatization of birds, "the soil's native products are scorned by the same current children of the country who raise animals and cultivate cosmopolitan plants without realizing that the native products of the land will represent an inexhaustible source of wealth."[54] Cosmopolitanism was no longer the first or best path for development; Argentina and Argentines were valuable resources in their own right.

Onelli expressed similar sentiments, and over time began to focus more intently on acclimatizing Argentine animals across the country's climates. His numerous experiments with the *ñandú* (rhea, a South American flightless bird distantly related to the African ostrich and the Australian emu) demonstrate this increasing interest in the local. Originally, Onelli had attempted to acclimatize the African ostrich, valuable for its plumage, which is showier than the rhea's and thus better suited to millinery and the fashion industry in general. The *Anales de la Sociedad Rural Argentina* had especially promoted his conclusions about the exotic bird, publishing an image of the African ostrich with the title "Spurned rural riches" and the caption "African Ostrich (from the Palermo Zoological Garden). Easily acclimatized and prolific, especially in alfalfa fields."[55] Four years later, however, Onelli would argue in the same journal that as much as he wished to create wealth in the countryside, there was no reason to "discount the poor native [rhea] who, like a good Creole, knows his needs and lives perfectly in his homeland's environment, paying, if you will, his modest tribute."[56] The *ñandú*'s advantages over the African ostrich included its ability to mate with multiple females. Additionally, the clay soil of Buenos Aires province stuck to African ostriches' feathers, making the bird scratch itself, ruin its feathers, and thus eliminate any potentially higher value.[57] The initial fervor for the

African ostrich had turned out to be unjustified. Onelli viewed acclimatization of exotic animals as part of his job as zoo director but recommended that "we never allow ourselves to be deceived . . . by the mirage of the exotic," instead turning an eye to the preservation and propagation of useful native species whenever possible.[58] Once again, the zoo's animals seemed to prove the point that cosmopolitanism's and immigration's benefits had limits, and nothing good could come from an excessive preference for the foreign at the expense of the Creole.

In fact, the *ñandú* offered the opportunity to go against the tide of immigrants and imports, instead exporting the best of the Río de la Plata region. Onelli argued that the Creole ostrich was, in fact, the darling of "French ornithological and acclimatization societies and of all the farm owners in that country," who believed it could be used for food and fashion.[59] Argentines may not have seen the value of the native, but Europe—always ahead of the game—had. Lehmann-Nitsche reached a similar conclusion in an article on the acclimatization of Argentine birds (the elegant crested tinamou and spotted nothura) in Germany. Keeping the human parallel at the nexus of this reversal of center and periphery, Lehmann-Nitsche concluded that if the experiments continued to go well, soon Europe would give the birds "citizenship papers like those the imported Asian pheasant already has."[60] Instead of a peripheral country happily accepting European leftovers in the form of immigrants and culture, the success of Argentine species acclimatized in Europe proved the worth of Argentina, its fauna, and by extension, its people.

These debates over animal acclimatization and exploitation, published largely in technical journals, used the language of immigration to understand the processes of adapting foreign species to the Argentine territory. Given the social context of the time and the explicit parallels the scientists drew with human immigrants, they can also be read as meditations on the relative value of human immigrants and locals. Some immigrants were advantageous, bringing biological and economic benefits to the nation. This was particularly true when they (humans and sheep) could be persuaded to populate the provinces outside of Buenos Aires. Change was essential to this success. Immigrants could not succeed in their new country without adapting to the local context. While climate and diet were the two most important factors for animal acclimatization, for humans this was extended to the customs, laws, language, and culture of their new home. Acclimatization discourse also provided a further way to address immigration through scientific language and empirical observation, as well as advocating for a heavy, controlling hand in the process. Reindeer could not just be dropped in Argentina without guidance; similarly, human immigrants needed be shaped by scientific hands, providing support for the hygienic projects analyzed in Chapter 1.

THEY'RE HERE! THE ZOO AS ASSIMILATORY PROJECT

Thus far, I have demonstrated how ostensibly scientific concerns at the zoo—the composition of collections and acclimatization projects—surpassed the zoological and engaged with broader national concerns over identity, immigration, and *argentinidad*. These debates reached a portion of the public through scientific journals and the zoo's own materials, but the audience was limited due to the technical focus. Nonetheless, the analogy between human immigrants and exotic zoo animals was also developed extensively in the popular press for audiences spanning all ages. In these representations, species, or, more commonly, individual animals were personified. Like Mimí the giraffe, these animals embodied the immigrant experience, but also enabled commentary on pressing questions such as citizenship, assimilation, and national identity.

Animals could not be expected to participate in the rights and obligations of citizenship, nor make autonomous decisions about their place of residence or national affiliation, yet in the press they served as a vehicle for exploring these questions. One issue was whether immigrants should naturalize. Law 346, passed on October 8, 1869, laid out the requirements for naturalization: one had to be eighteen, have resided in the country for two years or be married to an Argentine woman, participate in public service, or help with colonization or industry.[61] While some, particularly Socialist Party leadership, encouraged immigrants to naturalize so they could vote, not all saw the benefit of doing so. Many immigrants planned to return to their home countries and did not want to denounce that citizenship.[62] As such, in 1914, only .4 percent of the total population were naturalized citizens.[63] In the broadly appealing magazine *Caras y Caretas*, the example of African wildlife, who would never be able to take advantage of the benefits of citizenship, questioned why humans would want to go through the process:

> What interest can there be in a Gnu or a Warthog becoming legal citizens? And why should they forget their Black Continent? And the fact that the Siberian deer now share water, air, light, and American life is not a reason for them to want to share their antlers with the deer of the land. If those animals could speak, if they could speak to say the consecrated phrases, "my children's

homeland, *porca América*, the country that hosts us, *pays des sauvages*." But they do not go beyond neighing or growling![64]

Similarly, a reader wrote to *Caras y Caretas* to say, "there's no reason to be so demanding. Is it, perhaps, essential that the leopard recently sent from London to our Zoological Garden take citizenship?"[65] In these examples, exotic animals are a vehicle for expressing the continued affective and political distance between adult immigrants and Argentina.

Questions of biogeography, citizenship, and belonging were much more clearcut when discussing immigrants' children. The many references to future dynasties in discussions of acclimatization made clear that it was a generative process. Full success was defined by reproduction, not individual adaptation. Even if adult immigrants assimilated, they would never be fully Argentine. Their babies, born in the territory, were. Indeed, as early as 1893, Holmberg was assigning citizenship to animals born at the zoo based on the definition established in the 1853 Constitution: animals and people born in the Argentine territory were Argentine, although only men had political rights.[66] In the lists of prepared hides for the zoo's small museum, he separates "Species' Homeland" and "Specimen's Homeland" (fig. 3.4) As such, a Bengal tiger is listed with a species homeland of "Tropical Asia," but having been born in Argentina, its individual homeland is "B. Ayres (J.Z.)."[67] Onelli continued this usage in his inventories and guides, listing Buenos Aires as the origin of new hybrids and babies born at the zoo.

The territorial definition of citizenship was also seen in the popular press. In the same article in *Caras y Caretas* where he doubted nationalization's value for a wildebeest or warthog, Onelli wrote, "The solution would be easy for the porteño children of bears, camels, and hippopotamuses: but that has already been resolved. We have a Creole tiger, son of Bengal tigers, much more tiger than his parents, and that is without the famous melting pot of racial fusion."[68] As in Holmberg's inventory, a tiger born in Buenos Aires was a Creole tiger. The same magazine reported the birth of the "First South American 'canna' antelope," as well as the birth of an "Argentine" camel.[69] Over and over, an animal's biogeography—even if on the plaque outside its enclosure—was subsumed by Argentine nationality if born in the territory. The message was clear: the children of immigrants were Argentine and should identify as such.

These new "citizens" were touted as Argentine success stories. This dynamic is best illustrated by the birth of an Asian elephant in 1906, the first such animal in the world to be born in captivity and survive. Encouraging elephants to reproduce in captivity was incredibly difficult and many

JARDIN ZOOLÓGICO DE BUENOS AYRES.

MAMÍFEROS.

CUEROS ARMADOS EXISTENTES EN EL MUSEO DEL JARDIN
EL DIA 31 DE DICIEMBRE DE 1892.

(INVENTARIO).

Núm. de ejemplares	NOMBRE TÉCNICO	NOMBRE VULGAR	PATRIA DE LA ESPECIE	PATRIA DEL *(ó de los)* EJEMPLAR	MACHOS	HEMBRAS
	MONOS.	*(él ó la)*				
	Catarrinos.					
1	Cercopithecus mona.	Mono de cara azul.	Africa.	—	1	—
	Platirrinos.					
2	Cebus fatuellus.	Caí, Monito comun.	Sud-América trop. et circa.	Paraguay.	2	—
	LEMURIDOS.					
2	Lemur macaco.	Maki-vari.	Madagascar.	—	2	—
	CARNÍVOROS.					
	Digitigrados.					
	Felinos					
2	Felis leo (jóvenes).	Leon, L. de melena.	Africa.	B. Ayres (J. Z.)	2	—
2	Felis tigris (jóvenes).	Tigre real.	Asia tropical.	B. Ayres (J. Z.)	1	1
1	Felis onça («Viejo»).	Jaguar, Yaguareté, &.	Ambas Américas.	B. Ayres.	1	—
1	Felis onça (cachorro).	» »	» »	B. Ayres (J. Z.)	1	—
1	Felis eyra.	Eyrá.	S.-A.: R. Arg., &.	Chaco.	1	—
1	Felis Geoffroyi.	Gato montés.	» »	Rep. Argentina.	—	1
	Plantígrados.					
	Ursinos, etc.					
1	Ursus arctos (jóven).	Oso pardo.	Europa, &.	B. Ayres (J. Z.)	1	—
1	Nasua socialis.	Coati.	Sud-América.	Chaco.	1	—
15				Al frente....	13	2

FIGURE 3.4. Inventory of hides in the Buenos Aires Zoo's museum elaborated by Eduardo L. Holmberg. The species' homeland and specimen's homeland are different when the specimen was born at the zoo. Eduardo L. Holmberg, "Jardín Zoológico de Buenos Ayres: Mamíferos. Cueros armados existentes en el museo del Jardín el día 31 de diciembre de 1892," *Revista del Jardín Zoológico de Buenos Aires* 1, no. 4 (April 15, 1893): 116. Image courtesy of the Academia Nacional de la Historia de la República Argentina.

prestigious zoos had already failed. The successful cultivation of a desirable exotic, under the careful guidance of the zookeeper, augured well for Argentine science, zoology, and thus modern culture, while also reflecting Buenos Aires's ability to incorporate foreign elements, generally for good and progress. As Onelli wrote, the baby "was welcomed with affection, feted, frequently visited because her birth would tell those abroad of the goodness of Buenos Aires's climate, the fine culture of the metropolis that had obtained such a rare product and showed appreciation for it."[70] Her image circulated internationally, and she became a powerful symbol of Argentine pride and power.

Equally potent was her representation as the ideal product of immigration. From her name to representations in the papers, the infant's Argentineness and citizenship status were overdetermined. This representation reiterated that children of immigrants—even pachyderms—were Argentines and should behave as such. The baby elephant was given the name Phúa Victoria Porteña. "Phúa" harkened back to her parents' origin, while she was given "Victoria" and "Porteña" because her "birth signifies a triumph for Buenos Aires."[71] The foreign part of her heritage is offset by two markers of Argentineness. This change is especially notable when contrasted with the names of her parents, Sayán and Nayán (alternately spelled Siam, Saiam, and Naián). Even foreign visitors were aware of her status as national symbol. US journalist Arthur Ruhl wrote (misgendering her) that the baby was extremely popular, and "as he was the first elephant who could claim to be an Argentine, he was very important indeed."[72]

The nationalist reading of the baby elephant circulated widely. *Caras y Caretas* crowed, "At four in the morning, the elephant in our zoo happily gave birth to a baby elephant, authentically native to our country. Authentically. Therefore, we should be proud. And once proud, we should ask that the corresponding identity card [*libreta cívica*] be issued to our compatriot the little elephant."[73] The repetition of "authentically," paired with "native" and "compatriot," argues for the importance of place of birth over any particular racial or cultural markers for identification. Born in Buenos Aires, the baby elephant was a citizen and should be allowed to get an identity card and exercise the rights and duties associated with it (although as a female, these would have been notably curtailed compared to those of a male elephant). Upon her death, the children's magazine *PBT* reiterated her nationality, calling her "fully Argentine" ("argentina a carta cabal").[74] Human children of immigrants, it seemed to suggest, should also identify that way.

Finally, exotic animals could also be used to express anxieties about immigration, crime, and disorder. Concerns about anarchist immigrants attacking the government and immigrants as vectors of crime led to the 1902 passage of the Ley de Residencia, which established that the government could expel any foreigner convicted of crime or whose behavior compromised national security or disrupted the public order.[75] In the popular press, animals were used to explain and demonstrate the consequences of these laws. In 1904, a gnu assaulted Onelli and made off with half his jacket (and almost took his pants!), a direct attack on establishment authorities. A *Caras y Caretas* article discussing the incident explicitly tied the gnu's crime to his being an immigrant, labeling him one of the "recently arrived" who "acquire morbid revolutionary curiosity as soon as they start walking from the dock."[76] The cartoon used anthropomorphism to explain animal behavior, but also reminded readers of the danger of certain immigrant groups and their susceptibility to revolutionary rhetoric and political activism.

The image of the gnu as dangerous immigrant also appeared that year in the magazine *El Gladiador*. A cartoon depicted a group of animals as politicians meeting in the Jardín Zoológico to debate whether there should be an official homage for Ignacio Albarracín, president of the Sociedad Argentina Protectora de Animales, on his birthday. The representatives, including a polar bear, a rhinoceros, giraffes, camels, and monkeys, argue vociferously, being repeatedly called to order by the congressional president, one "Terrible Gnu" (see fig. 2.3). During the debate, a dromedary responds to the president's attempts to censure him by noting "word on the street suggests that the animal that presides over us is a dangerous foreigner," evidently referring to the gnu's attack on Onelli. Making the political tie even clearer, just before this exchange the polar bear had previously announced that discussion of the "Ley de Residencia" was outside of their jurisdiction.[77] Within the metaphorical animal world of the zoo, Onelli was the public order. In these middle-class magazines, the gnu's attack on Onelli—perfectly normal from a zoological perspective—is thus rendered an attack against the nation. The gnu was a bad, dangerous immigrant and thus in need of control.

The physical space of the zoo inscribed this metaphorical narrative of immigration and assimilation on the physical bodies of zoo animals and visitors. Animal immigrants were foreigners and part of their value was that very exoticness. At the zoo, they had their own themed living environments, which often resembled the human architecture of their homelands (or at least what Argentine architects and zoo directors thought architecture in their homelands should look like), and plaques listing their place of origin.

In other ways, however, they assimilated to Argentina, and these changes were noted and publicized to visitors. For example, Holmberg wrote of the hippopotamus, "numerous people have made observations about the dietary regime to which it is subjected, and they are all correct in stating that it is not its natural diet in the wild. It is given lettuce, cabbages, bread, bran, potatoes and dry grass."[78] Similarly, the 1920–1921 guide announced that the hippo in liberty "eats only green grass and aquatic plants. Delicate water lilies, the innocent lotus are the usual diet of this rude pachyderm: in slavery, they are fed with succulent potato pastries, bread, carrots and barley."[79]

They also adapted to the climate: Onelli noted that animals whose coats changed in color or texture with the season eventually molted in sync with Buenos Aires's seasons, even when they were opposite those of their homelands. Their children born in Buenos Aires did so too.[80] Similarly, Oskar and Jane, two highlander cows from Scotland, never molted in their place of origin, but did in Buenos Aires's warmer climate.[81] Migrants were predisposed to transformation, able to shift even biological processes to suit the environmental conditions of their new homeland. Adaptability was the name of the game. These examples, as well as the metaphorical representation of zoo animals in the popular press, presented a positive narrative to the public of immigrants assimilating to the conditions of their new environment. While they might never be citizens themselves, they were assimilated to a certain degree, and their children, born in Buenos Aires, were "authentically Argentine."

In the face of increasing immigration, racial or ethnic identifications with the nation began to break down, requiring the creation of new nationalisms that depended on cultural or territorial definitions. Animals were ideal for representing this new relationship with the nation, for while species borders made it impossible for an elephant to ever be racially Argentine, or even to form a common physical or biological relationship with other residents of the zoo, they could still be represented as embodying an Argentine spirit, and thus be "legitimately Argentine" despite remaining elephants. If species lines could be crossed in *argentinidad*, then race and ethnicity posed no threat. At the same time, the animals' bodies and the zoo space itself created a space for cultural identification, nationalist education, and civic (and often national) pride.

Onelli stressed the importance of assimilation by singling out Jewish immigrants as a negative example. In a 1917 speech at the Centro Juventud Israelita, he drew on well-trod stereotypes to address questions of ethnicity, immigration, and nationalism. He explained that in modern Argentina, religion was no longer the dividing social factor, but rather race and ethnicity, two terms he used interchangeably.[82] He then accused his Jewish audience of always choosing to identify as Jews over any national identity, a pattern he claimed did not exist among Catholics, Protestants, or atheists.[83] Echoing long-standing antisemitic tropes, he feared Jews were not "melded with the nation as a whole."[84] Indeed, he complained, Argentina had long opened its doors to all, including Jews, eschewing racially based hate, but the Jewish youth repaid this openness by continuing to separate themselves from "the rest of the Argentine race." "It is not enough for you to be Argentine: you are and want to be Israeli youth," he censured.[85]

As the immigrant animal must adapt to its new environment and its children be publicly lauded as Argentine, he exhorted his audience to stop creating ethnic institutions and assimilate: "Be Jews, Muslims, Semites, Arians, Mongolians; be it in your inner core and in your traditional intimate manifestations if you want. But I beg you, for the good and peace of future generations, give up that title in the development of culture and in the performance of the great activities you are called to carry out in this part of the world: be only Argentine."[86] In the same way that species could represent some far away Africa but be Argentine, Onelli believed immigrant groups who came to the zoo could maintain a certain ethnic identity at home while being Argentine in public. To further clarify, he insisted that he "admire[d] in you, young people, all your race's beautiful qualities and, for this reason, wish[ed] that when you are blended on American soil, you be fully Argentines, without clarifying adjectives."[87] In a 1918 article in the Jewish journal *Vida Nuestra*, he again argued that Jews possessed "an enormous potential for assimilation and adaptation" and that in contrast to Europe, in Argentina the wall between Jews and non-Jews was still low enough to breach.[88] Nonetheless, he lamented that they continued to insist on their separate identity and Zionist dreams, refusing to see that the real promised land was in Argentina.[89] In the speech and the article, Onelli stressed the importance of assimilation while also insisting on a hierarchy of immigrants that derived from their willingness to abandon their past. Importantly, as I will show in the next section, Onelli did not apply this criterion uniformly. At the same time as he scolded Jewish immigrants for clinging to their ethnic identity, he established the Latin roots of other immigrants as the basis of the new "Argentine" race.

At the Buenos Aires Zoo, animal enclosures physically demonstrated for visitors the messages of immigrants' foreignness reigned in, controlled, and wielded to produce benefits for the nation. While the animals' themed enclosures reminded visitors of their homelands, these enclosures were circumscribed by the fences of the Jardín Zoológico de Buenos Aires, bringing them under municipal control. Unlike other zoos organized taxonomically or based on region of animal origin, the Buenos Aires Zoo was organized around aesthetic principles, meaning that domestic and exotic species were mixed. The only geographical constant was the subsuming of different foreign regions to the municipal space of the zoo. And while the institution was a city project, the strength of the capital's association with the nation and the dearth of regional zoos until the late 1910s allowed it to also stand in for the nation. As Catalan journalist Santiago Rusiñol wrote, the zoo was a melting pot for animals and for the humans that came to visit, speaking different languages and professing different ethnicities.[90] But the category of "pensionista" or "visitor" lumped them together in a municipally owned, nationally identified space, rendering them, if not citizens, then Argentines in space and identity. Animal bodies built the nation.

A short vignette from *PBT* reinforces the nationalizing of the zoo's boundaries, as well as the zoo's continued focus on children as the prime target of assimilatory efforts. The Buenos Aires Zoo and State-sponsored public education were closely tied together as children accompanied by their teacher were always granted free entry to the institution. Many schools took advantage of this, with *El Monitor de Educación Común* reporting on and encouraging such field trips. During this time period, schools also increasingly espoused patriotic education to encourage unity among the diverse population.[91] In this context, in January 1912, Antonio Prado published the article "The Alsatian and the Zoo Orator," in which he described two "singular types" found in Buenos Aires. The second type was a *criollo* who sat outside of the zoo entrance every day that field trips would be visiting. When school groups approached:

> With the teachers' prior permission, the man delivers patriotic speeches, which he knows by heart, about the historical records of San Martín, Belgrano, Moreno, Rivadavia, Sarmiento, etc. As the teacher trails off, he artistically delivers strong impressions, modulating his voice and little by little becoming emotional, until he sheds intense and copious tears. The children, faced with the spectacle of the crying patriot, share the speaker's emotion, and he, with visible effort and understanding that the opportune moment has arrived, requests a few coins for the poor proud Argentine! [*pobre argentino de corazón*].[92]

Although the "patriot" was not the zoo's direct employee, his actions and Prado's description highlight the patriotic, assimilatory function of the zoo. The orator's speeches outside of the zoo walls connected the municipal institution to a greater national identity, making the animals inside not just *porteños*, but also Argentines. Additionally, children brought to the zoo by their teachers would pass by and be explicitly primed to think patriotically as they entered the grounds. This preparation would allow them to fully appreciate the lessons in immigration, assimilation, and nationalism produced by the organization of the zoo, representations of the animals, and printed materials such as guidebooks.

IN DEFENSE OF THE RAZA LATINA

While many zoos in the Global North emerged as displays of imperialism, the Buenos Aires Zoo and its animals emerged during Onelli's tenure as a site for resistance to that same power.[93] In the early twentieth century, the United States had developed a particular vision of pan-Americanism that looked to box out European influence, establishing the US as the hegemonic power in the region instead.[94] Rather than a meeting of equals, these projects projected the US as a guiding light that would diplomatically, financially, and culturally guide the newer Latin American nations. Many Latin American artists and intellectuals met these projects with concern and developed alternative proposals. As Mark Petersen has shown, in the Southern Cone, many "were willing to work within pan-Americanism, so long as it fit their own policy agendas and ideas of regionalism."[95] Indeed, Argentines Roque Sáenz Peña and Vicente Quesada articulated political paradigms to counteract US power with strong regional organizations that would cement Argentine hegemony and leadership, including Sáenz Peña's proposed Liga Latinoamericana (Latin American League). Others, such as Manuel Gálvez and Octavio Bunge, ideated literary contestations and educational programs that positively reevaluated Spain's role in the Americas and promoted a "raza latina" against an "Anglo" race.[96] Importantly, the idea of the *raza latina* "[drew] heavily on shared historical and sociological foundations, rather than on biological ones."[97] In this context, Hispanoamérica (or Latinoamérica) emerged as a means to "formulate and affirm a common identity vis-à-vis the United States."[98]

Once again, commentators found the zoo useful for articulating concerns about the issues of the day. In 1906, against the backdrop of US secretary of state Elihu Root's visit to South America for the Third International Conference of American States, *Caras y Caretas* published a vignette in which one of the zoo's elephants was in a bad mood until his keeper whispered in his ear, "Long live the Drago Doctrine!"[99] One of Argentina's goals for the conference was to convince Root to express support for the Drago

Doctrine, originally articulated in 1902 to advocate against foreign intervention in the region due to the failure to repay national debts.[100] A month ahead of the conference, the elephant gave voice to this desire. Two years later, the magazine *PBT* examined Argentina's relationship to the US in "El Zoológico social," a series of panels that explained common human types in Buenos Aires by comparing them to zoo animals. In one, a tiger is depicted as a "formidable patriot," in the next, another big cat, labeled "Felis Washingtonius," is said to have supported Root's visit to Argentina in 1906 and is called an "unbearable Pan-American."[101] The tiger stands proud and elegantly dressed, while *Felis Washingtonius* is unclothed and looks embarrassed. The value of the two positions is crystalized through the other animals' reactions to them: in the panel with the tiger, a foxlike creature removes his hat and bows. In contrast, two monkeys and another big cat point and laugh at *Felis Washingtonius*. Excessive fealty to the United States was worthy of mockery, especially in contrast to those who acted as true patriots.

Onelli shared these concerns. In his writings, he frequently used animals to juxtapose the Argentine way of life and that of non-Latin Europe and the United States, finding the former preferable. A 1906 reflection on bee species provided the metaphorical material for this project. In it, Onelli announced that the contrasts in the customs of European honeybees and the camuatí, "its American counterpart," reflect those of "the human hives that develop in the same American environment."[102] The European honeybee is clean, orderly, and hardworking, but consumed by "the devouring fever of progress, of those who want to make a fortune, of those who store riches upon riches." In contrast, the Creole bee works just enough to "live peacefully and independently," takes a siesta, and ultimately "lives free."[103] Onelli lambasted the excessive materialism he associated with English-speaking populations and linked Argentines with a more moderate approach that allowed space for the spiritual. Over time, he refined the materialist pole from Anglo populations in general to the United States specifically, in parallel with the shift from British to US influence in the region. In fact, these formulations spiked in 1913, perhaps not coincidentally the year that Theodore Roosevelt himself visited Argentina and the Buenos Aires Zoo.[104]

Onelli's next rhetorical turn was to identify Argentina with a broader *raza latina*, superior to the *raza anglosajona* that defined the United States. In accordance with the metaphorical representations of animals we saw above, he largely defined the races in cultural, not biological, terms. He associated Latin-ness with the humanities, classicism, philosophy, metaphysics, art, poetry, and a sense of beauty.[105] In contrast, the Anglo race was, like the European honeybee before, associated with an excessive love for the material at the expense of the spiritual. Anglo efficiency and democracy might improve bad things, he argued, but they also reduced excellent things to a boring mediocrity.[106] Again, these ideas resonate with the work of Martí and other modernists writing around the turn of the century. The stressing of Latinness, abundant in Argentina but less so in countries such as Bolivia or Peru, also positioned Onelli's adopted country as the natural leader of a non-US-led pan-American movement, dovetailing with contemporary calls for Argentine hegemony.

Onelli's rhetoric grew more intense over the years. In a speech at the Club Francés on the eve of Bastille Day, he again criticized the United States for excessive materialism. He also argued that the East was not a viable alternative, for it only provided "Marxist poisons." Instead, he found fraternity among the French, Spanish, Italians, and Argentines who shared a god, "the cultured, civil, human, supernatural, sublime God of the Latin race who has shaped us in his image for the same lofty and noble ideals."[107] These ideals, he continued, were not work and money, but rather spiritual and immaterial. The various Latin countries in Europe thus had a duty to promote the *raza latina* in the Americas and defend it from the advances of both extreme US capitalism and eastern Marxism. Notably, Onelli's discourse remained firmly rooted in Rodó's dichotomy between material North America and spiritual Latin America even while many of his contemporaries began to adopt political, commercial, and military frameworks for understanding American relations.[108]

In addition to his speeches, Onelli used architecture and design on the zoo grounds to promote Latin culture. While the animals' enclosures often reflected the human culture of their places of origin, Greek- and Roman-inspired art decorated the zoo grounds. These included statutes such as *La Cigale*, *Diana cazadora*, and *La cabra amaltea*, all installed between 1909 and 1911, as well as an Italian fountain.[109] The 1920–1921 zoo guidebook featured those sculptures, and works purchased in Europe by artist and director of the Museo Nacional de Bellas Artes Eduardo Schiaffino. Among them were *El Eco*, sculpted from travertine, which Onelli called "the monumental stone of ancient Rome," and a sundial featuring the Latin words "horas non numero nisi serenas" (I count only the hours that are serene).[110] These artistic decisions reflect Onelli's personal identity as an immigrant from Rome, as well as the return to Greek and Roman culture called for by José Martí and José Enrique Rodó and explored by modernist writers like Rubén Darío.[111]

While the Palermo Zoo site was bigger and more visited, Parque Patricios was where Onelli had near total control over the design from conception to opening. There, he allowed his preference for classicism to run wild (fig. 3.5).

EL PARQUE ZOOLÓGICO DEL SUR

1. El Templo de Vesta, para aves, en el nuevo jardín zoológico del Parque de los Patricios. — 2. La Loba romana que caracteriza el ambiente. — 3. El intendente y el director del Zoológico señor Onelli, en el acto inaugural. — 4, 5 y 6. Vistas de la concurrencia.

FIGURE 3.5. Photographs of the inauguration of the Parque Patricios satellite zoo. The Roman-inspired architecture and statuary reflected Clemente Onelli's heritage and his belief that Argentina was a Latin country. "El Parque Zoológico del Sur," *PBT* 11, no. 517 (October 24, 1914). Image courtesy of the Ibero-American Institute–Prussian Cultural Heritage Foundation.

As described in Chapter 1, the "Southern Zoo" boasted buildings modeled after Roman and Greek temples. Latin phrases could be found around the grounds, including on the new restaurant, which proclaimed "Aut bibas aut abeas" (either drink or leave), encouraging the consumption of tea, coffee, or lemonade. The dining room was pure "Pompeian," including murals of pagan Rome.[112] Numerous photos of the Patricios site were published in the 1920–1921 guide to the Palermo Zoo. Tellingly, none of the photos were of animals. They were all depictions of the Greco-Roman architecture.

In his writing about Parque Patricios, Onelli also drew explicit parallels between Roman culture and the Argentine past and present, depicting the nation as the heir to Roman glory. The buildings housing bathrooms and ostriches were roofed with "colonial tiles that are absolutely identical to those built by the Romans," connecting the Spanish colonial past with a particularly Latin identity.[113] A statue of the Roman wolf that dominated the park further proclaimed the intertwining of past and present: under Argentine skies, the "emblem of eternal Rome" transformed, "something like a new life that she symbolizes in the very new environment of this new Latin race."[114] And reflecting more recent ties to the *raza latina* were the advertisements that filled the zoo guide's pages and thus paid for its printing. Among the products represented and specifically noted as Italian are the Compañía Italo-Argentina de Seguros Generales, Bosca y Cía sparkling Italian wines, Minghetti and Attenuati cigarettes from "royal Italy," Pineus Pirelli tires, "Vov" zabaglione, Chiribiri cars from Turin, the bed manufacturer La Italo Americana, and sparkplugs used by the Italian army, navy, and several Italian factories.[115]

If the first part of Onelli's anti-US project at the zoo was a defense of the *raza latina*, a second part was the cultivation of a unified Latin identity for the region, creating a block that would be powerful enough to resist the US. In 1913, *Caras y Caretas* published a special issue dedicated to "the Latin race in both Americas."[116] The articles in the issue highlighted the beauty of the region, praised the conquistadores, and extolled the virtues of Spain and the Latin race in general. Many also highlighted the similarities across the region, such as "Tipos populares," which began "Throughout South America, from the Gulf of Mexico to the Strait of Magellan, the Creole type is as brave as weapons themselves, long-suffering and sober as the best soldier in the world."[117] The issue also included an article by Roque Sáenz Peña promoting the idea of a "Latin American league" that would politically unify the Latin race.[118]

Onelli contributed to the issue with an article titled "La fauna de América Latina." In it, he applauded the movement to seek out "a beautiful utopia of brotherhood of the peoples that constitute this race," suggesting it could be revelatory for both the Old and New Worlds. However, he insisted that humans from Argentina knew very little about humans from Mexico (for example), making it useless to "look for the connecting links between humans." Animals were the answer. For Onelli, it was fauna that "gives us a community of types that could be a rhetorical reason to tighten the screws and demonstrate that under the American sky, whether that of the Northern Hemisphere's temperate zone, the hot tropics, or the frozen southern regions, nature often reveals itself as uniform and produces types and wonders that it has denied to other regions." Across America, nature was coherent, boasting creatures consistent from north to south and unique to the region. He then enumerated a variety of animals that existed from Mexico to Tierra del Fuego. These included hummingbirds, big cats, armadillos, parrots, flamingos, monkeys, bats, crocodiles, condors, and foxes. For each animal, he gave examples of the diverse countries in which they lived, as well as a few words about their appearance or behavior, which he rendered the same across great distances.[119] Despite existing in the region prior to colonization or national borders, animals become united as "latino-americanos" in a contestatory pan-Americanism that emphasized a Latin American identity separate from Anglo America or Europe.

Onelli's most explicit defenses of Argentina, Latin America, and the Latin race against the encroaching United States came in 1924, the year that he died. Founded that same year by V. D. Caranci, the magazine *Idea Latina* was intended for Italian immigrants and distributed through Italian social clubs and mutual aid organizations.[120] *Idea Latina*'s explicit goal was to "achieve the reunion of two worlds, that of Latin Europe and that of the South American countries," in line with Onelli's expressed philosophy.[121] In an early issue, the zoo director wrote about the Ministerio de Agricultura de Argentina's edition of Virgil's *Geórgicas* published in issue 215 of its Circular (February 1924). The ministry's introduction to the pamphlet said it was for the "recreation and solace of the Argentine farmer." In his article, Onelli expressed surprise that a "somewhat Yankee-ized ministry" would make a "Latin gesture," a spiritual echo of Petrarch in "the prosaic modern democratic life." These statements are interesting, for Onelli explicitly notes that the US is not just a future threat but had already influenced Argentine institutions. The publishing of Virgil was thus a pleasant surprise, and Onelli expressed his excitement for the *raza latina* to continue and spread.[122]

Finally, just weeks before he died Onelli gave a speech at the YMCA, a US-based organization. Invited to speak about animals in his role as zoo director, he instead warned against the "invasion" of US money and

Protestantism. Beginning with Theodore Roosevelt's statement to Onelli's friend and colleague Francisco P. Moreno that it would be difficult for the United States to absorb the "Latin countries" as long as they remained Catholic, Onelli asserted that US pan-Americanism was an imperialist movement, and Protestant missionaries and institutions such as the YMCA were tools of that effort. Effectively, he believed that this "peaceful, convincing, and educational penetration" of the YMCA and other Protestant organizations, as well as North American banking and the diffusion of the *Boletín Panamericano*, were clearing the way for the formal annexation of South American states, already happening in Panamá and Puerto Rico.[123] He explicitly makes the equivalency in the phrase "the great Pan-American operation (conquest . . .)."[124] In the end, he accuses the YMCA of being "the noblest part of that propaganda forced by subconscious imperialism" and assertively challenges them in the belly of the beast: "But rest assured, we of Latin race and Catholic birth believe in the homeland of the earth with the traits and pure sentimentalism of our dreaming race, which, as the French thinkers say, does not want to be the Christian bastion of modern financial Judaism."[125] In 1939, that speech would be unearthed, and a transcript published by Enrique Osés in his nationalist newspaper *Crisol*.[126] In that article, Osés calls Onelli's vision prophetic; it would also live on in Roberto Orofino Púrpora's 1951 text *Por la unidad de Hispano América*.[127]

From its beginning, the Buenos Aires Zoo was a place to develop municipal and national pride, as well as to promote assimilatory policies for the many immigrants arriving in the Río de la Plata. Holmberg traced the first contours of this project, but it was the second director, Clemente Onelli, who really fostered it. Rejecting excessive cosmopolitanism, encouraging immigrants to assimilate, showcasing Roman art and architecture, and explicitly challenging the US's role in Argentina were all facets of the same urge to protect a Creole identity associated with the *raza latina*.

Onelli himself was the ultimate example of the project he promoted. Born in Rome, he arrived in Argentina in 1888. Although he continued to be proud of his Roman heritage (as seen in his defense of the *raza latina*), Onelli embraced Argentine culture and life, and self-identified as Argentine on zoo employment documents.[128] At his death, Gustavo Adolfo Albornoz highlighted this double identity in an obituary published in the *Boletín Sarmiento*. The homage begins by asking, "was don Clemente Onelli Italian? Yes, he was Italian in terms of his pride in his race, in his birth, in terms of that Roman haughtiness found in the midst of his unparalleled straightforwardness, he was Italian in terms of that truly Italian talent, in love with all things beautiful, and at the same time always playful, humorous, a man of the world." But the same paragraph ends with "Yes, don Clemente Onelli was Argentine!," an identity that Albornoz attributes to Onelli's dedication to developing "Creole industries, provincial industries."[129] Over the course of the paragraph, Onelli's identity slides from "he was Italian" to "he was Argentine." His assimilation—or acclimatization, as he might have called it—was so great that his future biographer, Diego del Pino, titled the book, *Clemente Onelli: El más criollo de los tanos* (Clemente Onelli: The most Creole of the Tanos [Argentine slang for Italians]).

Like his *pensionistas*, Onelli could not give up his ethnic background: he had been born in Rome. Nonetheless, like the giraffe, he acclimated. And as he exhorted the young Jewish students at the Centro Juventud Israelita, he embraced the language, traditions, and customs of his adopted country, becoming not Italian-Argentine, but Argentine "a secas," with no qualifying adjective. He then dedicated himself to defending the *raza latina*, his common bond with others in Argentina, to the world.

INTERLUDE

Between 1880 and 1910, 36 percent of immigrants to Argentina returned to their countries of origin, whether as intentional seasonal migrant workers (called *golondrinas*, or swallows) or due to financial failure, homesickness, or myriad other reasons.[1] This fact alone highlights one of the biggest differences between zoo specimens and human immigrants: if the animals failed to *hacer la América*, they could not choose to go home. The only way out was death.

The Buenos Aires Zoo's programs and even the name "Zoological Garden" pointed toward growth, the enhancement of human and animal lives. Nonetheless, death permeated the grounds. The zoo was constantly trying to scrape together enough horses to feed the *fieras* and, as we have seen, how to sacrifice them hygienically became a point of contention between the zoo, animal rights activists, and the zoo's neighbors. Despite Onelli's best acclimatization efforts, many species could not survive in Buenos Aires's climate or on the diets available to them at the zoo. Others were victims of negligence or even murder. The zoo's location on the Avenida Sarmiento meant its animals were exposed to the noises of protests and parades, often featuring fireworks and noisemakers. To give just one example of the toll of these events, the 1910 celebrations for the Exposición Industrial (Industrial Exposition) caused the injury of a cassowary and an African ostrich, as well as the deaths of another ostrich, a seal, and a three-month-old zebra.[2] Humans also intentionally killed animals, including mandrills, birds, and snakes that were shot at or poisoned.[3]

Others, Onelli and the public suspected, could not emotionally adapt and died by suicide. Poor Mimí the giraffe's nostalgia was the cute subject of a poem, but family separation and captivity had tragic effects on the zoo's *pensionistas*. None reveal this more than the case of the orangutan Petronio. Petronio was captured in Borneo, sailed to Hamburg, Germany, and sold by Carl Hagenbeck to the Buenos Aires Zoo. He arrived in Argentina on October 2, 1907.[4] Local papers rushed to take his photo, with *La Nación* claiming that it was the first time an orangutan landed in America. Seemingly attracted by his repulsiveness, his mixture of "impulsive criminal and melancholic madman," Petronio particularly won the attention of the city's women.[5] *Caras y Caretas* took the opportunity to "interview" the orangutan, noting that he was a "monkey of such high lineage" and that they visited him in his "palace." According to the interview he liked philosophy, had a "poetic soul," and "striking an intellectual pose, told us about his nostalgia and his pain. He has suffered a lot." Supposedly, Petronio had left behind a wife and child in Borneo.[6] This representation followed a long global tradition of depicting apes, especially orangutans and chimpanzees, as sensitive and possessing humanlike emotions.[7]

As an adult orangutan, Petronio had little chance of thriving in captivity. Within days, he became seriously ill, refusing to eat and suffering from constipation. Confident in their newly formed field of coprology, Onelli and Dr. Ricardo Lynch tried to give Petronio laxatives so they could obtain his feces to study. Petronio repeatedly refused to take them until they developed a differently flavored base for the medicine. Microscopic examination revealed no bad bacteria, just fat molecules, likely from his diet of rice soaked in milk. Onelli felt removing milk from Petronio's diet would help him, but the orangutan refused to eat the rice without milk and would not try fruits, vegetables, or any other products.[8] He continued to suffer from constipation. Dr. Lynch prescribed two medicines. One was a mixture of physostigmine salicylate (which treats delayed gastric emptying and belladonna poisoning), podophyllotoxin (a laxative that also treats parasitic worm infections), and belladonna, also known as deadly nightshade (for stomach cramping and gastrointestinal distress). The other contained belladonna, cocaine, and opium.[9] These medicines appeared to help for bit, but by December it was clear that Petronio was dying.[10]

While Lynch worked to treat Petronio's physical ailments, Onelli worried that the true cause was Petronio's unhappiness with captivity. When Petronio arrived at the zoo, he had been given a room and his physical needs had been taken care of. Nonetheless, he continued to hide in the back of his cage in the dark and refused to eat until "sunk by so much hopeless misery, he took on the resigned look of hopeless despair" (fig. I.4). Onelli imagined that Petronio was missing his "lost companion" that he had left behind with

EL ORANGUTAN ENFERMO

FIGURE I.4. Petronio, sick and sad. Clemente Onelli, "Nuevas adquisiciones," *Revista del Jardín Zoológico de Buenos Aires. Época II* 3, no. 12 (December 1907): 358.

his son in her arms and wanted to return to his role as "a loving father" with no human name. The ultimate cause of his illness was therefore "immense nostalgia, the profound moral evil that has eaten away at his physique."[11] Onelli even believed Petronio had asked for his freedom by touching his cage's lock while looking his human captor in the eyes. Onelli pretended not to understand and Petronio sank back in misery. He concluded, "Oh, man! Savage, civilized, ignorant, or wise, he is always evil: thus our poor Petronio must rightly think."[12]

When Petronio finally appeared for the public again on January 2, 1908, everyone wanted to see him. *La Nación* published two large images and a note claiming that the orangutan was much improved.[13] Despite their optimism, Petronio died four days later. His body was exhibited in his cage and visitors paraded by to pay their respects.[14] *La Nación* published a eulogy written by Onelli, in which he described the ape holding his hand while dying, "the hand of the white man, of the superior race that had defeated him, martyred him, destroyed his life."[15] He received 152 sympathy cards, including one from a little girl with a copy of Anatole France's poem, "La mort du singe." After the viewing, his brain was given to Christofredo Jakob, his autopsy was published in the *Revista del Jardín Zoológico de Buenos*

Aires, and his skeleton was sent to the Museo Nacional.[16] A true party for scholars, indeed.

Petronio's necropsy gave a diagnosis of chronic gastroenterocolitis, but Onelli and the press felt suicide a more likely explanation, arguing that the orangutan's nostalgia and profound sadness in captivity made him choose to starve to death.[17] Petronio's slow death was not unique, as zoo animals around the world have been documented electing to die.[18] Onelli's final words about Petronio are worth reprinting in full, as they clearly reveal the orangutan's wretchedness, as well as the tension between Onelli's sympathy for the animal and his conviction that the zoo was necessary for the improvement of human life:

> I knew a monkey, Petronio, the zoo's magnificent orangutan, hunted in the full vigor of his life and who, visibly terrified by the darkest nostalgia for his lost freedom, in just over a month, slowly allowed himself to die of starvation. Not tempted by the first fruits, nor by water itself in the great heat of summer, when the torture of thirst was already drying out his poor anemic and wasted lips, he refreshed his dry mouth for a moment and, consistent with the slow martyrdom that he had committed himself too, threw it away after a long swish. Poor monkey! How I understood your searching gaze, which so clearly asked me to open the locks of your confinement! Poor monkey, when you lay down again, resigned, I looked away, ashamed of your slow death in which I was an involuntary accomplice! Your memory, your slow suicide, today, five years later, still gnaws at my conscience, that conscience that cannot be reassured by the justification of the high duties of a jailer for public security, which forced me to watch you die.[19]

Indeed, Onelli frequently struggled with his role, calling himself a "jailer," "executioner," and "he who cuts off my prisoners' sweet idylls." With these words, he recognized that his *pensionistas* were not immigrants, not aristocrats, but prisoners forced from their homes to live in difficult conditions.

Despite these scruples, however, Onelli continued to acquire animals for the zoo. He justified his actions by proclaiming that the zoo's specimens had a mission to fulfill: "They are here so that the child and the ignorant learn to know them; here so that the psychologist can try to rip off the veil of their nebulous mentality, and thus be able to better understand them; here so that the pathologist can better discover the origin of diseases, study them, correct them, and heal them for the good of the animals and the human race." In this clearly anthropocentric view, the benefits they would provide for humanity were superior to the costs to any individual animal. If this argument failed to persuade readers, Onelli also suggested that zoo life might not actually be so bad, despite his deeply felt sympathy for Petronio. In his experience, "80 percent of the pensioners of a Zoario prefer their current quiet life to the eventful and dangerous life of the desert or the countryside." Animals that escaped from the zoo quickly came back, so perhaps they did not love freedom the way humans said they did.[20] Finally, although he was a jailer, he claimed to be a benevolent one, "the prison guard who allows himself to be moved, who allows violations of the regulations to occur, who allows and facilitates certain comforts not customary in a rigorous prison."[21] Onelli clearly cared for at least some of his animals and felt bad about their captivity. Nonetheless, his resistance to disrupting existing hierarchies existed for species as well as for social class. One could feel bad for animals or the poor and work to improve their lives, but that did not justify turning power over to them.

Petronio's death also spurred reflections on animal intelligence. Onelli mused that if animals died by suicide, it might be some of the best evidence for their intelligence. Survival was at the root of all instinct, so to choose death was to intentionally overcome the drive to endure.[22] An animal that chose to die had examined its situation and reasoned to the conclusion that death was preferable. Journalist Marcus Humo had similar thoughts. He argued that Petronio was more rational than the men who got drunk and made resolutions on New Years Eve, only to feel sad and suicidal by the spring when they failed. Petronio "resolved the question of new life in an unexpected way. He understood that new life must be, at a minimum, a better life than the old one: and without beating around the bush, which is no small virtue in a monkey, he immediately passed on to a better life." Trapped in captivity, Petronio could not bear another year of the same and made the logical decision to end his existence.[23]

Petronio's experience demonstrates yet another way that animals frustrated intentions and expectations as well as the mixed emotions *porteños* had about animal captivity. Holmberg and Onelli always worried about their budgets: exotic animals were expensive to purchase, needed round the clock care, and often ate large quantities of food. The animals were supposed to repay those costs by bringing visitors to the zoo, turning a profit, and educating the masses. By choosing to die, animals made all those plans impossible, denying eager visitors, scientists, and financial administrators what they wanted. Zoo animals have limited space to actively rebel. They can bite, they can refuse to act, they can try to escape, but the biggest act of resistance might just be dying.

Finally, in the early twentieth century as in the present, the Buenos Aires Zoo was a place where humans came face-to-face with their conflictive relationship with animals. The reactions to Petronio's life and death also show that awareness of captive zoo animals' misery is not a present-day invention. City residents were thrilled to see an orangutan up close, but nearly every article written about Petronio highlighted the tragedies of his life. They described his pain in great detail and explicitly made the connection between his sadness and his captivity. Many even appeared to accept suicide as the rational choice in his situation. In Petronio's haunted eyes, they saw a glimpse of themself as cruel tormentors, felt a sympathy that might temporarily have broken down the walls we build between humans and animals. We can only wonder how long that moment of insight lasted, if it was gone by the time they reached the next enclosure and returned to marveling at the beasts in their cages.

CHAPTER 4

THE NATIONAL FAMILY GOES TO THE ZOO

Reproduction, Motherhood, and Feminist Ostriches

At approximately 80 kilos, she was the stork's biggest delivery yet. Born on February 23, 1906, Phúa Victoria Porteña, daughter of Nayán and Sayán, was the first elephant in the world to be born in captivity and survive. Her birth drove Buenos Aires crazy: 22,650 visitors crowded around the Hindu temple enclosure on a single day, eager for a glimpse of the zoo's newest resident.[1] Within a few months, the director estimated that 20 percent of the city population had gone to see her.[2] Newspapers and magazines celebrated her birth with poems, cartoons, and many, many photos.

Phúa's case is extraordinary, but reproduction was always big business at the Buenos Aires Zoo, providing financial, reputational, and scientific benefits. Purchasing an elephant from a trader like Carl Hagenbeck cost thousands of dollars; a baby was free and would even create profit by drawing in new visitors willing to pay the ten-cent entrance fee to gape at him or her. Even more common species like Patagonian maras were useful, as the zoo could trade extras with other institutions for novel species. Additionally, as many animals do not reproduce or infrequently reproduce in captivity, success in overcoming these challenges represented scientific progress. Finally, given the biological burden of reproduction, the arrival of the stork demonstrated that zoo animals were in passably good health, shorthand for a zoo's quality. Baby animals could thus be proof of a city's modernity and a source of pride. As Onelli boasted of Phúa Victoria Porteña, "her birth would tell those abroad of the goodness of Buenos Aires's climate, the fine culture of the metropolis that had obtained such a rare product and showed appreciation for it."[3]

The zoo and its animals also performed a social function in codifying a maternalist, heterosexual national culture where sex was for reproduction and the State had an important role to play overseeing pairings. With a rapidly growing and increasingly heterogeneous population, many Argentines fretted about the fate of future generations. Prostitution, the dissolution of the family, single motherhood, and abandoned or sickly children were perceived as signs of social and racial disorder that could halt or even undo progress. In the words of medical doctor and former national deputy Eliseo Canton, children's "vigor and future were equally important to society and to the country's government" and thus demanded action.[4] Given the strong association between the zoo and childhood, it is perhaps no surprise that elites found it to be a productive space for these interventions.

In the decades around the turn of the century, zoo stories, the representation of employees to the public, and programming provided the models and materials for bringing about the desired "healthy and robust children, capable of later becoming useful men for society and the *patria* [country]."[5] These efforts spanned a narrative arc from pairing to parenting. First, they enforced heteronormativity and insisted that marriage's primary purpose was the production of future generations. Parallel to this discourse were limited musings on the possibility of scientific and State control of those pairings. Second, the zoo offered gendered models of behavior, particularly in relation to child rearing. These paradigms were naturalized and ahistoricized through human-animal comparisons. Third, the zoo and its satellite sites contributed to the material conditions necessary for defending traditional motherhood in the face of modernization and industrialization. Consistent across these three types of interventions was a fundamentally maternalist approach to women and their role in the nation. Finally, the zoo also offered an exception that proved the rules in the form of *ñandúes*, also known as rheas (*Rhea americana*). In the wild and in captivity, the flightless birds share incubation duties and chick care across genders. In Buenos Aires, zoo administrators and observers seized upon this "aberration" as a potent tool for critiquing feminist demands that destabilized motherhood as women's natural and most sacred duty.

MARRIAGE AND MATING

Heterosexuality was a structuring principle for the Buenos Aires Zoo's collection. According to the institution's original regulations, when purchasing animals there should always be an effort to "complete" a set of specimens, acquiring a male if the zoo already had a female, or vice versa.[6] Similarly, the rules prohibited selling a "pieza única," defined as a solitary specimen or a male or female from a single pair. Keeping male/female couples permitted the study of gendered physiology and behaviors at the zoo. And as Noah's Ark's two by two suggests, it also ensured the ability to produce future generations, many of which director Clemente Onelli recorded in a notebook cheekily titled "Genesis."[7]

The *Revista del Jardín Zoológico* and zoo guidebooks focused intensely on presenting animals living in monogamous, heterosexual pairs. Those that remained in stable pairings, such as the *chajá* (southern screamer, *Chauna torquata*) were compared to "civilized races" and depicted positively.[8] The African ostrich was lauded as an "austere monogamist," who was angered when Onelli introduced females into his hut, transforming it into "an oriental harem."[9] When Onelli describes nonmonogamous animal relationships, they are usually depicted in critical language that posits them as unpleasant or even hellish for the individuals in them.[10] These observations were also inherently raced, associating alternate forms of sexuality with human others, including Indigenous people. These links had long histories in Argentine understandings of Indigenous people, with men in particular defined as lascivious, sexually violent, and perverted.[11] Interestingly, whereas in earlier nineteenth-century texts Indigenous sexuality was dehumanized through animal metaphor (men are wolves, for example), in these texts, stereotypes of Indigenous sexual behavior are given as the analogy for understanding animal sexuality. Out of the exuberant diversity of animal sexual behaviors and kinship units, the zoo imposed a singular narrative of monogamous relationships as the only "civilized" option.

The physical spaces of the zoo also promoted heterosexuality. Whenever possible, enclosures featured family units of animals, including a male, female, and their young. Zoo stories described these groupings with the language and ideals of human marriage. Obtaining specimens for breeding programs involved proposing: "The Zoo Director has asked for the front paw of the beautiful and hairy North American bison for his new pensioner, the buffalo recently arrived from Washington."[12] Specimens were then described as being "legitimately married" or as a "legitimate married couple."[13] What did it mean for elephants or monkeys to be legitimately married? It meant nothing to them, of course, but everything to the society that observed them and found parallels with human life. For Argentines after 1889, weddings had to be carried out by a civil servant and registered with the State to be legitimate.[14] That zoo marriages were witnessed and announced by Onelli reiterated that pairings needed to be sanctioned by authorities, be they the zoo administration or the civil registry, thus strengthening national control over citizens. "Marrying off" zoo specimens also inscribed any sexual behavior within marriage, making animal action align with one particular form of human morality.

The green spaces and infrastructure of the Buenos Aires Zoo encouraged humans to emulate the animals and form State-approved heterosexual pairs. As Catriona Mortimer-Sandilands has shown for US and European urban parks, the benches and gazebos that punctuated the lawns encouraged heterosexual courtship.[15] In Buenos Aires, couples of all social classes used the Buenos Aires Zoo as a backdrop to their flirtations. Wealthier couples promenaded on weekdays, while on Sundays, waves of working-class men and women put on their finest clothes and met among the animal enclosures. Onelli believed his zoo provided a virtuous backdrop for courtship. Unlike taverns or cinemas, the utopic zoo had no dark corners for scandalous activities; everything was exposed to the light. His article "La Arcadia Porteña" demonstrated the exact contours of the zoo's promotion of a chaste, Christian, heterosexual national culture. In it, an archetypical working-class couple flirts on benches surrounded by flowers, in clear view of all passersby. The girl sits demurely at a safe distance from her suitor, knees together and hidden by her big skirt. The boy speaks of getting the money needed to take her "to church." The zoo is romantic in a respectable way that contrasts with the (perceived) promiscuity of working-class spaces outside the park. Like the zoo stories about animal couples, "La Arcadia Porteña" upholds the traditional marriage structure and solidifies the State's role in sanctioning it, concluding: "the popular Sunday excursion to the Zoo contributes with its light, its air, its open shadows, to increasing the civil registry's lists."[16]

Zoo stories also demonstrated that the ultimate goal of marriage was reproduction. Stories of animal weddings are quickly followed by hopes for children, and the creatures were often referred to as mother and father, not wife and husband. For example, in the 1906–1907 Zoo Guide, a photo of the elephants Nayán and Sayán is labeled simply as "The parents of Phúa Victoria Porteña."[17] Their most important role in the zoo was producing the first baby elephant to be born and survive in captivity; this action subsumed their individual identities. The local press also gleefully reported on new arrivals at the zoo, always in positive terms (fig. 4.1). In 1911, *PBT* ran a full-page

El acontecimiento zoo-cial de la semana

Eureka!... Honradamente no nos atrevemos á afirmar que el antiestético cuadrúpedo que nos honra con su nacimiento, sea el primer "camello" que haya visto la luz en la República Argentina. Sin embargo, debemos aceptar el advenimiento del pequeño rumiante como un feliz presagio de prosperidad nacional. El orgullo patriótico debe estar satisfecho. Bajo nuestro cielo, ya no sólo prosperan elefantitas como Phua y reinan monos como Cónsul III, sino que también nacen burros, zorros y camellos... Aunque el hecho no nos ha sorprendido, queremos dedicarle un sitio reservado para consagrar gráficamente el más notable acontecimiento zoo-cial de la semana. El señor Onelli, como director de los habitantes del Jardín Zoológico, y padre adoptivo de ellos, merece una congratulación por el notable aumento de la población.

El señor Onelli, contemplando paternalmente á su nueva pensionista

FIGURE 4.1. The birth of the first Argentine camel was a sign of progress and source of national pride. "El acontecimiento zoo-cial de la semana," *Caras y Caretas* 9, no. 404 (June 30, 1906). Image courtesy of Cornell University Libraries.

spread on a baby camel named "Polo Sur," including photographs, information about its species, its birth weight, and its market value.[18] The magazine *Sherlock Holmes* announced the same birth, gushing that the mother "was very celebrated, we could even say applauded, for the splendid event" and the father was rewarded "with an extraordinary meal, and with a spontaneous and enthusiastic 'well done' from Mr. Onelli."[19] *Sherlock Holmes* was a magazine dedicated to police narratives and stories of crime; the birth of the baby ungulate must have been exciting enough to depart from its generic norms. The exhibit of zoo families and reports about them tightly linked marriage and reproduction, while also codifying heterosexual families "by placing animal specimens and their human observers along the deep time line of natural history."[20] Shared by zoo animals and humans, heterosexual, reproductive relations were "normal" and thus desirable.

The ubiquity of zoo birth narratives and the collective celebrations of them in the press also inscribed reproduction within a communal nation-building project. For humans, demographic growth had been linked to Argentine nation building since politician and intellectual Juan Bautista

Alberdi's 1852 declaration that "to govern is to populate." Late nineteenth- and early twentieth-century discourses continued to depict reproduction as a simultaneously personal and collective endeavor.[21] At the zoo, the birth of an elephant or a camel increased the population of a beloved municipal institution and raised Argentina's scientific and cultural profile on a world-wide stage. Offspring thus belonged to the nation just as much as to their parents. Indeed, a 1913 commentator explicitly tied together humans, zoo animals, reproduction, and nation, asserting that elephants in Europe rarely survived in captivity, but "Here [in Buenos Aires] they grow and procreate, as if they know they are in a country where the first civic duty consists of increasing the population."[22] In this way, elephants modeled good citizenship for their human observers.

Although zoo babies were a public good to be watched, commented on, and celebrated, the reproductive process itself remained cloaked in secrecy. As Berlant and Warner so clearly articulate, "National heterosexuality is the mechanism by which a core national culture can be imagined as a sanitized space of sentimental feeling and immaculate behavior, a space of pure citizenship."[23] Sex is associated with privacy and intimacy, an act that takes place away from the public gaze.[24] The zoo archive operated on this same logic, for despite pages and pages of romanticized animal relationships and an unceasing attention to birth, actual sex is rarely depicted. When it does briefly appear, it is hidden under layers of euphemism (an "interesting ceremony") and the intimacy of the moment is stressed, such as when an American puma and an African leopard celebrated their marriage "at the bride's house, in the greatest privacy."[25] As metaphor for human reproduction, the State used resulting offspring to construct the future nation but the actual sex act had to remain private to perpetuate national heteronormativity.

CONTROL OF MARRIAGE AND REPRODUCTION

The zoo's scientific mandate and the captive nature of its animals meant that it was fertile ground for experimenting with more intense interventions around reproduction, the results of which had mixed applicability to humans. At the zoo, Onelli could not force animals to mate, but by putting them together in enclosures or separating them, he could exercise a degree of control that city elites could only hope to someday wield. For example, when Onelli wanted the Patagonian maras to reproduce, he had his workers place them in communal living groups.[26] When he worried about the male/female balance of the zoo's peacock population, too many males being a "sign of the beginning of degeneration," he had them cut off the least dominant males' tails, removing them from the mating pool and avoiding gendered conflict.[27] And getting deer to mate without the males injuring the females required "some trickery and a little surveillance."[28]

Holmberg, Onelli, and their employees took advantage of these circumstances to carry out numerous studies on inheritance. In 1904–1905, the zoo's first vet, M. González Herrera, bred two lions, Ruka (female) and Menelik (male). Ruka was very strong, with "good teats" and "proves to be a mother in everything and for everything." In contrast, Menelik was weak, cross-eyed, had very developed testicles, did not urinate forcefully, and had a weak roar.[29] González Herrera noted that in the wild, neither Menelik nor his children would have survived, for natural selection would triumph. But because he was in the zoo and the number of lions was limited, the uneugenic lion was given a chance at reproduction. González Herrera's concerns about the father's lack of fitness were substantiated when three of the four cubs died, a tragedy he attributed to the male's influence. As he concluded, the cub Lerdo "was destined to die, lymphatic, with a poor organization. He had inherited all his father's bad conditions and so could not live to be old." The only cub who survived, the vet noted, was "slender, flexible, majestic, he has inherited from Ruka all his race's magnificence."[30] While captivity could circumvent the struggle for life, González Herrera's experiment showed that biology was destiny. Creatures born to bad parents had few options for survival.

Onelli himself was particularly interested in Mendel's laws, recently rediscovered by Hugo de Vries and Carl Correns. The zoo's large collection of chickens (an animal slowly being phased out of European zoos) was particularly useful for investigating the inherence of dominant and recessive traits.[31] Onelli's experiences with poultry, as well as guinea pigs and the free cats that roamed the zoo, led him to write to Argentine naturalist and politician Ángel Gallardo with doubts about the universality of Mendelian principles. If what he found in his animals proved true for humans, it would be as if "a first-bred mulatto, born white enough like his Caucasian mother, had his complexion darken and his almost straight hair become kinky upon turning eighteen."[32] Gallardo wrote back arguing that the guidance provided by Mendel's laws was an improvement over the blind breeding to which they were previously limited.[33] Onelli's reluctance to accept Mendelian law wholeheartedly aligned with that of many of his contemporaries who were hesitant to accept biology as destiny because of its social-political implications. In a country with an Indigenous past (and present, although most did not see it that way) and immigrants from all over the world, a purely biological understanding of race was dangerous. Instead, many preferred a

Lamarckian-flavored evolution that allowed for the influence of (and control of) the environment.[34]

Onelli also believed that animals in the wild were perfectly eugenic, for, as he told the Sociedad Sarmiento, "among beasts, love is not a toy nor entertainment for young and old; it is the great eliminator of useless and weak lives; it is the god who is always an ally of and gives victory to the strong and healthy in order to keep the race whole." As a result, physical anomalies, psychopaths, the feeble, and the freaks were rare in the animal world. When they did exist, they were the result of human intervention or captivity, as seen in the case of Menelik and Ruka's lion cubs. Pure love "knows how to choose among the beasts those that best respond to its eternally equal, eternally eugenic, eternally perfect purpose."[35]

Onelli was more skeptical of human eugenics. The same year as Onelli's speech at the Sociedad Sarmiento, Argentine doctor Víctor Delfino was invited to the Second International Eugenics Congress as the only South American presenter. Although the Congress was delayed because of World War I, Delfino would go on to found the Argentine Eugenics Society with Gregorio Aráoz Alfaro, Joaquín V. González, and others, including statesman Estanislao Zeballos.[36] Onelli's experiences with zoo animals led him to conclude that efforts to implement eugenic marriage among humans would fail. He argued that human biology was highly analogous to animal biology, but social forces meant that "love will not be subjected to laboratories."[37] Class, money, and other social pressures interrupted the natural process, "perverting" love. Onelli's rejection of human eugenics is in notable contrast with some of his counterparts in other parts of the world, such as the United States, where the administration of the Bronx Zoo included founding members of the American Eugenics Society Madison Grant (secretary of the board of the New York Zoological Society from 1895 to 1925 and president of the society from 1925 to 1937) and Henry Fairfield Osborn (president of the New York Zoological Society from 1909 to 1925).[38]

Finally, the zoo contributed to the movement to scientifically control reproduction through collaboration with medical professionals. Onelli's tenure coincided with national efforts to medicalize reproduction, including the emergence of the fields of gynecology and obstetrics. The 1897 creation of an Escuela de Obstetricia at the Facultad de Ciencias Médicas de la Universidad de Buenos Aires separated female *parteras*, who historically had accompanied laboring women, from male doctors. The masculinized side of the profession was associated with science, logic, hygiene, and progress.[39] As seen in Chapter 1, Onelli believed that comparative anatomy and pathology had a lot to offer to various medical specialties; obstetrics was no exception. He took an interest in the field, approaching his gestating animals with a clinical gaze and delivering a speech at the Círculo Médico Argentino, "Obstetricia agreste y vislumbres treponémicas" (Rural obstetrics and treponomic glimpses), in 1920. He worked closely with Doctor Victor Widakowich, an Austrian-born gynecologist. Widakowich was the chair of Histology and Embryology in the Facultad de Ciencias Médicas at the Universidad de Buenos Aires and a "distinguished professor of the maternity ward" at the Hospital Ramos Mejía. He also served as the Buenos Aires Zoo's administrator and director of the laboratory in the 1910s. Together, Onelli and Widakowich worked on topics such as coprology, microbiology, and generally improving animal health. In addition to sharing knowledge, Widakowich was a physical go-between for the zoo and the city's medical institutions, delivering the genitals of deceased zoo animals to the director of the Obstetric Clinic for study. This cooperation, Onelli insisted, "is not just useful but also necessary for scholars to reach their full potential."[40] Although never dealing with human patients, the Buenos Aires Zoo played at least a small part in the medicalization of reproduction and paving the way for scientific interventions.

MOTHERS AND FATHERS

In addition to promoting heteronormativity and the scientific control of reproduction, representations of the zoo's collections and its human workers contributed to the maternalization of women, a process shared by politicians and activists across ideologies. Between 1880 and 1920, the increasing importance placed on children's well-being and the development of hygiene, eugenics, and puericulture crafted a vision of the Argentine family that was "nuclear, patriarchal, legitimized and legalized by laws." Within this family, the father was the bread winner, while the mother was a housewife and moral compass, dedicated to raising the "child who acquired capital importance for society, the State, and the 'race.'"[41] Motherhood was her natural and most important role.

In zoo stories, female animals were almost exclusively defined in terms of their maternity, making motherhood appear to exist before culture and uniformly across species. For example, the 1906–1907 guidebook included a picture of the lioness Ruka captioned, "the great-great-grandmother of our lions."[42] In contrast, the male lion Bruto was described as "the fiercest of our lions."[43] He was defined by his personality, particularly elements that confirm stereotypes of lions (they are fierce beasts) and men (they are strong). When Ruka died, she was again circumscribed within her reproductive function: "Ruka, the old lioness, the wife, mother, grandmother, great-great-grandmother of the loveliest lion collection in the world; Ruka, who could have been photographed in the middle of a group of forty-seven lions, eight generations that she has seen be born."[44] Her progeny were her legacy; no other characteristics or actions were worthy of remembrance. A cultural commentator also used the zoo to stress the singularity of women's role. They proposed that the young elephant, Phúa Victoria, died partly because she had learned that women were meant to be mothers and was unable to reconcile this fact with the knowledge that there were no possible mates for her in Buenos Aires. Unable to fulfill her mission on earth, the author suggested, she chose to expire.[45]

The association of women and motherhood was also incorporated into the visual vocabulary of the era. In the *Sherlock Holmes* article about the camel family, one photo depicts the baby camel nursing from its mother, an act often seen as the greatest expression of the mother-child bond and almost always biologically impossible for men. To further emphasize the mother-child dyad and the father's separate, independent existence, the male camel appears in a completely different frame.[46] Similarly, in texts spanning the infantile to the technical (*PBT*, *Revista del Jardín Zoológico de Buenos Aires*, *Anales de la Sociedad Rural Agentina*), the baby elephant, Phúa Victoria, was almost always pictured with her mother, Nayán. Her father, Sayán, either did not appear in the articles or was pictured alone in a separate photo. The few times that he appears in the same frame as Phúa Victoria and Nayán, the baby stands close to her mother with her father standing at a significant distance. Photos of the pachyderm mother and child even circulated as postcards, further spreading the maternalist perspective. That visitors saw the animals as models for humans is evident in the note one "Juan María" wrote over a copy of that postcard in 1906: "Darling: imitate Phúa, who does not go out for a walk without her mother."

The Argentine maternalist movement elevated maternal love above all else, asserting it was both natural and something that could be taught or encouraged, as it "domesticated women."[47] The extensive observations of animal behavior and psychology that Onelli published in the "Idiosincrasias individuales de los pensionistas del Jardín Zoológico" section of the *Revista del Jardín Zoológico* presented this love as instinctual. Onelli once claimed motherly love was so evident that there was no need to prove "the innate love the female has for her offspring," but he nonetheless spent the next two decades writing about it.[48] According to Onelli, the multitude of examples among zoo animals suggested that maternity existed prior to culture and was thus universal, timeless, and ideal. Indeed, even the female animals who had not been partnered nor reproduced were enlisted into motherhood and ensuring the continuation of the species. When the titi monkeys began dying of cold, the head zookeeper suggested "giving them an ordinary monkey as a nurse, she, like an old spinster, became deeply fond of her adopted children and sheltered and cared for them at all hours of the day and night."[49]

Onelli made this point explicitly in a speech to the Consejo Nacional de Mujeres (National Women's Council) in 1914. Founded by Cecilia Grierson, Albina Van Praet de Sala, and others to advocate for women, the organization split apart in 1910 when Grierson and the other "universitarias" left, leaving behind an upper-class, Catholic faction that prioritized charity over feminist activism. It was this more conservative group of women that invited Onelli to speak. In his remarks, he again insisted on the existence of a "maternal instinct well ingrained by the laws of nature," making motherhood women's supreme responsibility and right.[50] Onelli exhorted his audience to be like the elephants, who "knew" it was their civic duty to reproduce: "Mother, you know your mission: to have children so that your country's wealth, well-being, and culture develop."[51] His words perfectly encapsulate what Marisa Miranda argues is a defining feature of the time period: the reconceptualization of childbirth and rearing as a State issue rather than an individual or family one.[52]

As with eugenics, Onelli was convinced that the animal world demonstrated correctness in motherhood while humans were corrupted by social and economic forces. Although the period of maternal love was shorter in animals than in humans because of their accelerated development, animal mothers—driven purely by instinct—were often better mothers than humans. For the zoo director, proof of this deeper maternal love was found in comparing the animal protection agency, Sociedad Protectora de Animales–Sarmiento, with its famed human homologue, the Sociedad de Beneficencia. Both had built hospitals and financed ambulances, but the animal protectors never had reason to establish a "Foundling Home, those opulent buildings that are both the pride and shame of human civilization."[53] The laws of nature made women mothers; if they failed, it was not because they were biologically destined for something else but because society or their own egoism had disrupted natural processes.

Where were fathers in all of this? The consensus in early twentieth-century Buenos Aires was that the mother-child relation was natural and emotional, but the father-child relationship was social and financial. As Nari concisely states: "Women were mothers by instinct. Men by law."[54] Contemporary psychologists used beliefs about nonhuman animals to argue that men were not naturally caring fathers, but Onelli claimed his first-person behavioral studies of zoo animals found numerous examples of involved males, proving the existence of paternal instinct.[55] For example, a sambar deer (*Rusa unicolor*) at the zoo was "an exemplary father" who protected and lovingly groomed his offspring.[56] Father monkeys also seemed to feel affection for their children.[57] While Onelli admitted he mostly observed this instinct among monogamous animals, and especially mammals, even polygamous males exhibited "small attentions to their progeny" that could be considered signs of instinctive paternal love.[58] Paternal love might not be as powerful as maternal love, but his observations allowed him to sketch a limited role for men in child rearing that contradicted much of maternalist discourse.

The zoo's most lauded parents, the elephants Nayán and Sayán, served as object lessons for these two types of parental love. Despite never having given birth prior to Phúa Victoria, Nayán immediately knew what to do and proved herself to be good mother: her attentions were "transformed day by day into a loving and liberal protection, that trusted in the mental development of her offspring and only corrected or intervened when her old experience advised her to do so." In fact, her form of motherly love was superior, for "it was not exclusive, selfish, nor of that uselessly exuberant effectiveness of humans."[59] When Phúa Victoria fell ill at age three, Nayán became extremely upset, breaking the chain around her leg to be closer to her sick daughter. When the little elephant died, her mother refused to eat for days and was so distraught that zookeepers could not safely remove the baby's body.[60] The elephant's death inspired dozens of cartoons, poems, and reports, many of which anthropomorphized the pachyderm family and highlighted the emotional aspect of motherhood. In *PBT*'s "Visita de duelo," the narrator goes to the zoo to express his condolences to the grieving pair. Nayán is inconsolable and even crazy with grief. Sayán, who Onelli noted loved Phúa Victoria "in his own way," was upset but able to speak rationally.[61] As demonstrated by the elephants, zoo stories carved out a larger role for father figures and naturalized their paternal love, but always to a much lesser degree than that of mothers. A women's mission was to reproduce and raise children; fatherhood was just one aspect of a multidimensional masculinity in both animals and humans.

The gender roles described in these stories and images were also performed for visitors by the zoo's employees. All but one of the employees listed on the yearly budgets and payrolls between 1900 and 1924 were men, but there were several women who worked at the zoo in a more informal capacity. Ana Galli and María Ana Basualdo attended the women's restrooms around 1917, while Andrea Tello oversaw the "nursery," presumably the Templo de Vesta lactation space. These women were not paid with municipal funds but rather with money raised through the sale of eggs laid by the zoo's chickens.[62] This kept them from accessing any of the benefits of being on the payroll, such as a pension, and meant their employment was contingent on animal production. Textual and visual narratives reveal additional unnamed women caring for the zoo's infant animals. A female maid living in Onelli's house on the zoo grounds once opened the door to a visitor with a tiny puma in her arms, and some unidentifiable furry ball on her back.[63] Others appeared in photos in zoo guides and articles.

The women who worked at the zoo are uniformly depicted in maternal roles. In a 1912 *Caras y Caretas* article, one woman wears a collared shirt and skirt and carries a large pot. She is identified only as "the blond who makes food for the privileged animals." There are also two images of a woman wearing a dress and a ruffled white apron, gently washing a baby otter and holding two small caimans. The accompanying text presents her as "the charming caretaker of several generations of lions and tigers" (fig. 4.2).[64] Other women who worked with the zoo babies are referred to as a "selfless nurse," "our lion and tiger nurse," and the animals' "nodriza seca" (dry wet nurse, nanny).[65] They cooked, bottle-fed, petted, and otherwise supported the infants. Women, the representations imply, are so naturally maternal that their care crosses species lines. In fact, Onelli once boasted of possessing a monkey who was "breastfed by a [human] woman!"[66] The *Caras y Caretas* article also reflects the gendered medicalization of motherhood discussed previously: the female employees perform care work, while many of Onelli's daily activities were related to the science of reproduction. He calls the wildebeests over, then bends down to "check if there is swelling in the female's udders, a sure sign of family on the way." Later, he repeats the same movement with Torita, the female bison, and a zebra. The latter he praises specifically as a "fecund mother," before verifying her mammary glands were swollen.[67]

The one exception to this very strong pattern of gendered representation is María Esther Llambias, the only woman to appear on the official zoo payroll during the twenty years for which I have records. She began working at the zoo as a laborer in November 1917, earning the same salary as her male counterparts.[68] By January of the following year, she had been promoted to assistant poultry breeder ("2° avicultor") with a salary of 120 pesos, more than any laborer or zookeeper earned. This salary was also higher than that of her immediate predecessor, a man.[69] There are no further mentions of Llambias until March 15, 1919, when Onelli wrote to the mayor's secretary

ensarta sus gafas y con toda la flema escandinava, correspondiente también á la intelectualidad de un guarda-fieras, empieza tranquilamente á descifrar la carta. Mientras, el señor Onelli se zahuma con tabaco turco, fumando impaciente un cigarrillo tras de otro y dispara que se lo lleva el diablo cuando llega á saber que ha perdido un buen cuarto de hora para tener conocimiento de que un Mr. Petersen se ofrece como domador de leones.

A las nueve y veinte en Palermo; y, con una matraca que se bambolea hasta sobre el betún de la Avenida, estamos á las 9 y 40 en el criadero avícola de Belgrano: una mirada á las gallinas, otra torva al molino quijotesco que con sus aspas hacia el sudeste amenaza una creciente del río; á las 10 en el parque Saavedra, y á las 11 en el parque de los Patricios, frente á un lindísimo y alegre chalet hecho para lechería de cabras: ¡si hubiéramos sabido, cargábamos cuatro docenas de placas!

A las once y media otra vez en Palermo: grandes abrazos y grandes fiestas con el mono gibón «Cascarita»: después visita á los más íntimos, á los que hace cuidar en su casa.

La gentil cuidadora de varias generaciones de leones y tigres, nos presenta, como la madre de los Gracos, las pequeñas alhajas que ahora cuida: «Angelita», una lobita de río; «Mechita», la antílope más pequeña del mundo; seis escuerzos (éstos, por fin, no tienen nombre), á los que cuesta un

La rubia que hace la comida para los animales privilegiados

Haciendo la toilette al lobo familiar

La cuidadora de los pequeños yacarés

FIGURE 4.2. Some of the unnamed women who cared for animals at the Buenos Aires Zoo. "Jardín Zoológico: El día de un director," *Caras y Caretas* 15, no. 709 (May 4, 1912). Image courtesy of University of California, Berkeley libraries.

to pass on her request for a leave of absence, as well as that of the zookeeper Marcelo Sires. Both employees wished to absent themselves the following week, suggesting they may have been together.[70] Llambias stayed in the role until quitting in June 1919.[71] Llambias's employment breaks with the maternal roles other female zoo employees performed, but her uniqueness and the fact that she never appears in the guides or zoo stories only underscores the consistency of the maternalist representation of women working at the zoo.

With respect to gender, zookeeping was a somewhat ambiguous occupation. On one hand, the zoo was a scientific institution and science was still largely a man's profession. The zookeepers' work could also be dangerous, requiring them to interact with unpredictable beasts and use their masculine strength to subdue them. In public accounts and private requests to the municipality, Onelli frequently stressed this side of zoo work, even asking that his keepers be paid a bonus to compensate for those risks.[72] Additionally, as at the Bronx Zoo in the United States, the male employees wore uniforms that highlighted both their masculinity and their role as keepers of animals and order.[73] On the other hand, the employees' regular duties included care work that encompassed traditionally feminine activities. Keepers needed to feed and bathe the animals as well as clean their enclosures, all actions falling under the purview of the traditional housewife.[74] These moments were not hidden; the guides and popular press frequently featured photos of male keepers holding baby animals or caring for specimens. One particularly common visual trope was Onelli bottle-feeding animals, contrasting the masculine authority of his directorship with the feminine act of hand rearing an infant.[75]

A caricature of Onelli from November 1908 neatly illustrated that zoo fatherhood was more involved and took on traditionally feminine care work (fig. 4.3). The image, drawn by José María Cao Luaces, depicts Onelli in a white suit and a striped tie helping a small rhinoceros drink from a teacup. The civilizing force of Onelli is highlighted in the contrast between the dark rhino and Onelli's light suit and skin, as well as the juxtaposition of the jungle-like trees behind them and the china cup, saucer, and silver spoon Onelli has supplied. The image also emphasizes Onelli's hands-on approach to his *pensionistas*. The caption casts the relationship between human and animal specifically as a parental one: "El animalito está / perfectamente cuidado, / y Onelli ya le ha enseñando / a decir *papá* y *mamá*" ("The little animal is / perfectly cared for, / and Onelli has already taught it / to say *father* and *mother*").[76] In addition to infantilizing the animals and thus reinforcing human control over them, the representation of Onelli as father and mother perfectly highlighted the gender duality of zoo work.

The observations of zoo animal behavior that Onelli published in his "Idiosincrasias" (later reprinted in other magazines and independent books, boosting their reach) naturalized women's role as mothers, showing motherhood to be instinctual and all-consuming. Fathers, in contrast, did display fatherly love, but also retained other characteristics. Similarly, female zoo employees were exclusively maternal, while male employees engaged in stereotypically male and female activities without losing any of their masculine prestige. In this way, zoo stories painted a picture of women strongly associated with children and care, as well as fathers who took on some

FIGURE 4.3. Onelli was seen as both father and mother to his *pensionistas*. Cao, "Caricaturas contemporáneas: Clemente Onelli," *Caras y Caretas* 11, no. 529 (November 21, 1908). Image courtesy of the University of Miami libraries.

responsibility for their young while also existing in separate, masculine spheres.

PUERICULTURE: RAISING HEALTHY CHILDREN FOR THE NATION

For early twentieth-century technocrats, the scientific control of marriage and maternity could and should be extended to the proper rearing of children, known as puericulture. In the intrauterine and extrauterine environments (in the womb and after birth), mothers would play an outsized role in raising children that could contribute to the nation's future stability and development.[77] They thus needed the best information and scientific interventions to improve outcomes. Evidence of these efforts is illustrated by the 1899 publication, and many subsequent republications, of doctor Gregorio Aráoz Alfaro's *El libro de las madres: Pequeño tratado práctico de higiene del niño con indicaciones sobre el embarazo, parto y tratamiento de los accidentes* (The mothers' book: A small, practical treatise on children's hygiene with indications about pregnancy, birth, and the treatment of accidents). The text walked mothers through the best care of children from preconception hygiene (who should or should not reproduce), to what to eat during pregnancy, to how to bathe newborns. Other sections included feeding, child development, home hygiene, exercise, and moral education.[78]

Behind many of these topics was an interest in reducing infant mortality, considered one of the best indicators of "the population's sanitary and sociocultural conditions."[79] In Argentina in 1905, ten out of every one hundred live births ended in death, an alarmingly high rate compared to European countries.[80] In addition to general hygiene, reformers quickly settled on infant nutrition as a principal area for improvement. In the late nineteenth and early twentieth century, women across social classes breastfed, turned to wet nurses, and/or supplemented with other mammal milks, sometimes mixed with flour and/or sugar. *Higienistas* blamed the latter two for many of Argentina's medical and social ills. Wet nurses were accused of spreading moral and physical disease, while the mothers who employed them were "unnatural" for refusing to participate in what was seen as a key element of the mother-child bond.[81] Other infant food was nutritionally insufficient and had a high risk of contamination. Encouraging women to breastfeed their own children emerged as a key priority because, as Eliseo Canton wrote in 1913, "in addition to being a child's natural right, breastfeeding is health insurance that only she [a mother] can provide." Human milk was perfectly composed for human nutrition, free (if you discount women's time and freedom), and more aseptic than other animal milks and milk replacements, making it a lynchpin of efforts to reduce infant mortality. To encourage women to breastfeed, activists distributed flyers with a picture of a woman breastfeeding captioned "The child's most sacred right and the mother's most legitimate pride," and organizations such as the Liga Argentina contra la Tuberculosis organized breastfeeding contests.[82]

As with other elements of the *cuestión social*, the Buenos Aires Zoo collaborated in these efforts. Because it was a space that invited mothers and children, including young babies, to visit together, it was especially well suited. Physical reforms to the zoo grounds encouraged women to breastfeed during their visits. In 1910, the zoo inaugurated a new "Sala de lactancia para señoras" (nursing room for ladies) housed in a building that was supposed to be a replica of the Temple of Vesta (but was actually a replica of the Temple of Hercules). Located in the heart of the zoo, women could quickly stop to feed their babies in quiet privacy. Inside, the space was lined with the same white, square tiles that were used in many medical institutions of the city and the zoo's *matadero*, contextualizing infant feeding as a sterile, hygienic, and even scientific activity. A photo in the zoo guide reinforced this image. Rather than depict the women or babies who would use the nursing space, it shows a group of men proudly posing on the steps outside.[83] The Templo de Vesta nursery encouraged maternalization, the mother-child bond, and breastfeeding, while also revealing the increasing medicalization of puericulture and male control over female bodies and activities. This shift would only intensify in the late 1920s and throughout the 1930s with the rise of explicitly eugenic institutions that diminished women's autonomy in the fields of reproduction and childcare.[84]

The zoo's animals were also enlisted to promote breastfeeding and to shame human mothers into changing their behavior and placing child, and thus future nation, ahead of their own needs or desires. Pictures of baby zoo animals nursing abounded in the literature, naturalizing that modality of infant feeding. Commentators insisted that animals were better than humans, for they did not have wet nurses. Breastfeeding was a "superior pleasure" in which animal mothers voluntarily and happily participated.[85] Sources made clear that when zoo animals were bottle-fed, it was due to having been orphaned or because they were at risk of injury from their mothers. As in messaging for human mothers, alternatives to breastmilk were only acceptable in exceptional cases. A reflection on a female hippopotamus in *Caras y Caretas* summarized these attitudes toward breastfeeding: she is said to have "fulfilled, and continues to fulfill, her maternal duties with strict diligence." The article further wondered, "could many human mothers say the same, those who send the fruit of their womb to the foundling home, or those who give their children to a wet nurse for fear of ruining the lines of their body, sacrificing their child to the demands of the corset?"[86] Mothers who chose not to nurse were selfish and unnatural, and put their child at risk of death.

The zoo guides also disseminated information about infant and child health, both from the State and from for-profit industry. The 1908–1909 Zoo Guide included among its smallpox immunization information that it was safe for pregnant women and nursing mothers to be vaccinated, and doing so would protect their health and that of their children.[87] Ads in the guides also touted supplements and foods designed for mothers and children. The 1908–1909 guide further bolstered breastfeeding with an ad for "Bebé Malt Extract. Special food for mothers and children."[88] Among other effects, malt extract can increase the quantity of breastmilk produced by a lactating person, making it more likely that a woman will continue to breastfeed.[89] The 1906–1907 guide also included alternative food products such as "Phosphatine faliéres," calling it delicious and recommended for children after they weaned. The miracle product "facilitates teething, ensures good bone formation, prevents or stops defects in the child's development, prevents the diarrhea that is so common in children."[90] That same product was also advertised on a sign atop the tramway that circled in the grounds, maximizing the number of people who would see it. In the guides, these products were pitched in scientific terms and their printing by the municipal government suggested a certain degree of legitimacy and approval.

Perhaps the best example of the imbrication of the zoo and maternalist approaches to puericulture is the November 1917 Semana del Nene (Children's Week) sponsored by the Club de Madres, which was mostly composed of upper-class Argentine women. The week-long event aimed to reduce child mortality and increase health, particularly among the poorest families of the city. Events included the distribution of clothing and linens and a puericulture exposition with the participation of the Asistencia Pública, Sociedad de Beneficencia, Mothers' Canteens, Instituto Nacional de Bacteriológico, and private companies. There were also straightforward presentations on puericulture by local physicians, including Isabel Kaminsky's talk on "Alimentación del niño lactante" (Feeding the nursing child) and Alicia Moreau de Justo's speech "Bagaje de salud que los padres deben transmitir a sus hijos" (Health baggage that parents should pass on to their children).[91] These interventions targeted the mother-child dyad through the lenses of hygiene, science, and governmental programming with an eye to national productivity and well-being.

FIGURE 4.4. Poor women and their children visited the zoo for a tour and a snack during the first-ever Semana del Nene. Club de Madres, *Memoria de la 1a Semana del Nene organizada por el Club de Madres para contribuir a la disminución de la mortalidad infantil* (Buenos Aires: Wiebeck, Turtl, 1918), 29.

The Buenos Aires Zoo was invited to participate in this event that combined science and charity to combat infant mortality, demonstrating once again that reformers believed it had an important role to play in efforts to resolve the *cuestión social* and improve *porteño* society. The mothers and children in attendance were invited on Wednesday, November 21, to board tramways from all over the city that proclaimed in giant letters, "Buenos Aires wants mothers to be helped to raise healthy children" (fig. 4.4). The three hundred poor mothers and their offspring were taken to Palermo, where Onelli met them and gave them a tour of the zoo. It was many women's first visit. At 4 p.m., they had a snack in the Confitería del Águila consisting of milk from La Vascongada and cookies from Bagley y Cía.[92] There is no record of what was said during the tour but given the context of the visit and Onelli's speeches to women in other spaces, we can imagine it was at least partially devoted to discussing motherhood and childcare. It also served as an introduction to the importance and availability of fresh air in the city.

Finally, the Buenos Aires Zoo's approach was decidedly conservative when it came to women's education and work. Onelli's zoo welcomed schoolchildren with free admission and special programming and in many years more girls than boys visited on official fieldtrips. On at least one occasion in 1910, upper-class boys and girls were given the same scientific education, as students from the Liceo Nacional de Señoritas and the Colegio

Nacional were engaged in "practical studies preparing anatomical pieces, cutting tissue for microscopic examination."[93] Despite this moment of equality, Onelli viewed these studies as preparing the genders for different occupations. The boys' exposure at the zoo readied them for scientific careers. In contrast, in a speech to the same Liceo Nacional de Señoritas, Onelli implored the girls not to allow scientific learning to distract them from their true mission as wives and mothers. "Be, if you can, women of wisdom, doctors, and poets; but do not let those titles make you dizzy . . . above all, you must always be housewives, selfless companions like Leontina Poirier de Ameghino."[94] Like Poirer de Ameghino, wife of the famous Argentine paleontologist, they should aspire to "the most modest and sublime role, that of the self-sacrificing scholar's wife, withdrawn from the noises of the world, totally devoted to making the home peaceful for that man, for that genius who, struggling with the necessities of life and the infallible frictions of the arduous scientific task, with his magic pen, with the fossilized piece in his hands, powerfully and dazzlingly revived the life of millions of centuries ago."[95] If women had a role in science, it was to ensure that men's needs were taken care of so that they could focus on their intellectual pursuits.

Nonetheless, like many maternalist reformers, Onelli recognized that some women had to work to survive and fretted about how it would affect families. To resolve this problem, he worked with the municipality of Buenos Aires to establish a school for teaching traditional weaving skills to women. As Onelli and the mayor of Buenos Aires, José Luis Cantilo, argued in separate speeches to the public, the Escuela de Telares (Textile school) was a response to the modern world. One of the scourges of the present was the factory that "swallows childhood, youth and women, especially women, in the insatiable jaws of mechanical industrialization."[96] Given the association of the nuclear family with social stability, women's entry into mechanized out-of-home labor represented a threat to the nation itself.[97] Indeed, even feminist activists wanted to regulate and limit women's factory work.[98] In contrast, the hand weaving that would be taught at the school was a skill that could be done in the home, increasing the number of women who could do housework and rear children while also "contributing to meeting the ever-increasing needs of the modern era."[99] By teaching this "honest work in the home," Onelli argued, "the Commune fully enters into the correct way of interpreting modern feminism, taking care of the essential characteristic, to have fertile mothers and supervised and educated children."[100] Between October 3, 1921, and August 1, 1922, 263 girls and women attended the school; 160 of those received diplomas.[101]

The project also intersected with other foci developed at the zoo over the previous two decades. It was located at the Parque Patricios Zoo, ensuring that it would reach both working-class and immigrant populations. Furthermore, the weavings would be focused on traditional Argentine patterns, especially from the north, preserving and promoting the local.[102] As in other aspects of his revindication of nationalism, in textiles, Onelli framed the work as Creole and a resuscitation of the colonial era, titling his first study *Alfombras, tapices y tejidos criollos* (Creole carpets, tapestries, and fabrics) despite incorporating numerous Indigenous designs.[103] The press followed this example, describing the school as a workshop for colonial-era rugs and the weavers themselves as *criollos*.[104] The Escuela de Telares would thus allow poor women to work from home while teaching and disseminating the nation's cultural heritage, packaged to appropriate Indigenous heritage as the basis of a Creole identity that could stand strong against US incursions in the region.

UNNATURAL MOTHERS AND CRIMINAL ELEMENTS

That most representations of gender at the zoo were ultimately quite conservative can be seen when Onelli and others who wrote about the zoo addressed the emerging feminist movement in Argentina. Like mass print publications, hygiene, anarchism, socialism, and immigration, the early development of feminism overlapped with the first decades of the zoo's existence. In 1900, Cecilia Grierson and other upper-class women created the Consejo Nacional de Mujeres to link groups discussing women's issues.[105] In 1901, Elvira López finished her thesis "El movimiento feminista" (The feminist movement) and became one of the first women to graduate from the Universidad de Buenos Aires's Facultad de Filosofía y Letras (School of philosophy and letters). Shortly thereafter, López, her sister Ernestina, Grierson, Julieta Lanteri, Elvira Rawson de Dellepiane, and Sara Justo founded the Asociación de Mujeres Universitarias Argentinas (Association of Argentine University Women), which produced the Primer Congreso Femenino Internacional (First International Feminine Congress) during the Argentine Centennial in 1910. This event caused the so-called *universitarias* to split from the more conservative women in the Consejo Nacional de Mujeres.[106] As with contemporary social movements, there was no unified list of feminist demands, but rather a constellation of political programs: some advocated for women's educational rights, others aligned with the anarchist and socialist movements to push for protections for working-class women. At various points in time, these groups rallied around divorce, suffrage, women's civil rights, and other issues affecting Argentine women.

As he had done with socialism, Onelli established a distinction in his many speeches about women and to women's groups between a legitimate feminism and an extreme variant. In its natural form, Onelli argued, "feminism is a trend worthy of all respect; it is a perfection of Christ's idea that dignified women; it is an effort toward better customs; it is the dawn of a new era; it is the tendency toward the better half of the human race's participation in the rights and duties that until now, at least in theory, have been the exclusive monopoly of men."[107] Feminism was not ambition, but a call for women's entrance into the public sphere in line with mothers' natural care instinct. As in his public projects, women could and should be permitted a role in education and the world, so long as it did not interfere with their roles as "good wives and good mothers."[108] Anything else was a "absorbing tomboy-ism [mari-machismo]."[109] Onelli's opinion was not unique but instead echoed that of many Argentines who felt that rejection of motherhood meant a woman was monstrous, even criminal.[110]

Onelli viewed suffragettes as "unfortunate fanatics affected by a dangerous short-sightedness" who twisted natural, pure feminism into an artificial and immoral concept. The suffragette was "a bad bug . . . vermin that eats away at home and society."[111] Nature once again proved this truth, Onelli's colleague Fernando Lahille claimed, for the few species where the female was the same size or larger than the male were "the least differentiated beings," such as ticks and spiders. These revolting creatures proved that obtaining equality of the sexes "would be equivalent to lowering oneself to the most primitive and inferior forms of life."[112]

The children's magazine *PBT* also used the zoo's animals to critique more radical forms of feminism. In 1908, it published a cartoon called "El zoológico social," which associated various types of people with animals under the caption "It's so fun to see the animals that look like people!" One of the panels depicts "the sublimated goose (*anser sublimatum*), a whiny feminist who, not serving for anything that her sex should be useful for, wants to be a gander, which she does not serve for either, as is easy to understand." The image shows an aggressive looking goose in a dress and flowered hat, leaning toward a male-female pair of waterfowl as if to stridently express her opinion.[113] In 1912, the magazine profiled Karsavina, a five-year-old chimpanzee living at the Buenos Aires Zoo. Like other animals, she was married, and she was beautiful and modest while her husband Bertoldo was brutish. The author wrote that "when one thinks of the disparity between him and her, one feels inclined to applaud the violence of the English suffragettes so bravely led by Mrs. Pankhurst, even though their actions fall outside the Code," but then quickly dismissed the idea. The chimpanzee, more noble than some humans, "surely does not think like that, aspiring to feminine vindication not by those reprobate means, but by gentleness, according to the most modern System, which will surely have satisfactory results."[114] Like Onelli, *PBT* acknowledged that women might want some changes to their role in society, but rejected suffragism and activist forms of feminism in favor of "gentleness."

There was one example of this "artificial" feminism among higher species: the *ñandú*, or rhea (*Rhea americana*). Within a flock, rhea females lay their eggs in a communal nest built by a dominant male. The male incubates the eggs while the females leave and may even mate with other males. Rather than view the rhea as proof of gender diversity, Onelli and some contemporaries imbued their descriptions of the birds with harsh judgment and used them to explicitly critique more radical manifestations of feminism. The 1904 edition of the zoo guide explained to visitors that the male incubated the egg and "resignedly takes care of the new babies," while the female "runs through fields and goes about and gives herself to new loves."[115] The male sacrifices, hardly eats, and is gaunt and exhausted by the end of the incubation period; the female galivants. The Chilean educator and writer Rafael Sanhueza Lizardi also paid attention to these behaviors when observing the zoo's Creole ostriches in 1911. He compared the females to the Roman empress Valeria Messalina, known for her promiscuity. In contrast, he noted that the male—known in all other species as the one who sows his oats and leaves—"acts as the unnatural and clumsy mother, declares himself broody and incubates the eggs, from which he only separates for half an hour a day to eat." The poor creature suffers until the babies are old enough to be independent, at which point he finally "declares himself free again and recovers the strength lost in his long and painful devotion to home life."[116] What was considered natural and desirable for females was depicted as miserable, unnatural, and even injurious for males. As Onelli summarized in a speech to the Sociedad de Psicología, "look how masculinized feminism reduces the poor Creole ostrich!"[117]

Female rheas were such bad mothers that they presented a risk to their progeny. Onelli worried that the females, who he explicitly depicted as "carefree socialites, dedicated to the frivolities of useless excursions and eating scented bonbons" would, "in their little experience with motherly care, turn the future family into a colossal omelet."[118] This image aligned feminism with the literal destruction of the family and the future and was remarkably consistent over the years. Nearly two decades later, the 1920–1921 zoo guidebook called baby ostriches "stupid" because "they did not recognize the father that took care of them and ran around any other specimen of their species, probably their own mothers who, not accustomed to the essential work of all mothers, trampled them, dying at the feet of these unnatural mothers."[119] Anti-maternalist feminism could only lead to death and future instability driven by a declining population.

The reach of the dangerous feminist ostrich trope is demonstrated by the work of Georges Clemenceau. Between terms as the prime minister of France, Clemenceau visited Argentina for the 1910 Centenary and published his impressions in 1911. Clemenceau dedicated five pages of his text to his visit to the Buenos Aires Zoo, which clearly made an impression upon him (these reflections were also published in *The New York Times* and the Austrian *Neue Freie Presse*).[120] Onelli invited the former and future prime minister into his home on the zoo grounds and gave him a personalized tour of the most important animals. Commentary on gender and feminism marks Clemenceau's account of the institution. He notes that the penguins "would die of *spleen* if they weren't fed with a feeding tube like a simple English suffragette."[121] The comparison references the British suffrage hunger strike movement, begun in 1909 when Marion Wallace Dunlop was imprisoned in Holloway Prison and refused to eat. Other women adopted this tactic, leading prison guards to force feed them with liquid foods and rubber tubes.[122] Clemenceau takes a disgraceful moment in British history and facetiously compares it to the struggles of penguins forced into captivity far from home. He then turns to the rhea, echoing Onelli's assessment of their gender roles. He describes the male as a hard worker, dedicated to the care of the babies, while the female prefers to "loaf around" and "neglect her duties."[123] The ostrich is an unnatural mother, and the immediate proximity of this assessment to Clemenceau's reference to the English suffragettes suggests that they, too, go against nature. Indeed, Clemenceau's 1907 pamphlet "La 'Justice' du sexe fort" claimed "If the right to vote were given to women tomorrow, France would all of a sudden jump backwards into the Middle Ages."[124] For Clemenceau as Onelli, mothers that did not care for babies were unnatural and inherently anti-progress.

Two extreme examples of bad motherhood drove home to zoogoers that good motherhood was everything a woman could aspire to and that failing to achieve this ideal was a sign of a disordered and dangerous personality. The 1904 zoo guide included a picture of a lioness labeled "An exemplary mother." In it, she is surrounded by four lion cubs, standing over them as they also look out toward the camera (fig. 4.5). According to the text, this

FIGURE 4.5. The lioness Mogo was an exemplary mother. *Guía popular del Jardín Zoológico Municipal de Buenos Aires 1904* (Buenos Aires: A. Rothkopf, 1904).

UNA PESQUISA SENSACIONAL

HORRIBLES INFANTICIDIOS

DEL PALACIO Á LA CÁRCEL

ELLA JOVEN Y MIMADA

VERDADERO TIPO DE BELLEZA CAUCÁSICA

Vestía pellizas preciosas y comía miel

SU PALACIO ESTA EN LA AVENIDA ALVEAR

LA TENTACION DE UN MEFISTO

SE VEÍAN DIARIAMENTE EN EL ZOOLÓGICO — PECÓ — VOLVIÓ Á PECAR

CO IRREPARABLE Y SUS CONSECUENCIAS

EL SEDUCTOR LA ABANDONA — EL DESENLACE

DOS MELLIZOS DEL SEXO MASCULINO — SOFOCADOS EN LA ALCOBA

UNA INVÁLIDA DENUNCIA EL HECHO — EL TENORIO JUSTIFICA EL ALIBÍ

LA JUSTICIA HUMANA LO DEJA EN LIBERTAD

LA BELLA JOVEN EN CALABOZO — OTRO INFANTICIDIO EVITADO

NOTA — Estos títulos y las fotografías que publicamos á la vuelta, tomadas por nuestro reporter sobre el lugar del crimen, nos ahorran entrar en mayores detalles.

FIGURE 4.6. Drama at the zoo! A mother bear is accused of infanticide. "Una pesquisa sensacional," *Caras y Caretas* 15, no. 722 (August 3, 1912). Image courtesy of University of California, Berkeley libraries.

exemplary mother is fecund and "cares for her cubs tenderly." In contrast, another lioness "in the moment of motherhood becomes nervous, agitated, and gets annoyed with the public looking at her cubs, which she ends up devouring; for this reason, when childbirth arrives, we try to ensure that she is isolated and comfortable, protected by a wooden partition, and we try to take away her children as soon as they can be bottle-fed." In case it was not clear which were the preferred behaviors, the second lion is explicitly called "the bad mother."[125] This same phrase was repeated in the 1906–1907 guide.[126] The public who read the guide received an important object lesson about good and bad mothers, as well as the consequences for each. A good mother took care of her babies, while a bad mother did not and deserved to have them taken away and raised by others to ensure the future of the nation (or zoo).

A second murderous mother appeared in the 1912 article, "Una pesquisa sensacional: Horribles infanticidios" (A sensational investigation: horrible infanticides), about the death of bear twins at the zoo (fig. 4.6). Structured and illustrated like the crime reports becoming increasingly popular in magazines, the story told the tragic tale of an aristocratic woman living in a "mansion" on Avenida Alvear, one of the nicest streets in Buenos Aires. She was "a true Caucasian beauty" who wore furs and ate honey. Tempted by a "Mephisto" she saw at the zoo daily (so much for the chaste flirtations of "La Arcadia Porteña!"), she gave in to sin. Her seductor then abandoned her, pregnant. Until this point, she is not explicitly identified as a bear, although there are many clues. When authorities arrived, they found two male bears "suffocated" in the bedroom. Accompanying the text are slide-like images of the scene of the crime, the dead infants, the "filicide in prison" (ironically just her usual cage), and another bear "witness" that reported the crime.[127]

In both the bear and lion tragedies, the death of the infants results from the mother's mental state and her inaptitude for motherhood. The lioness is a "bad mother," too nervous and unfriendly. Ms. Osobich, the bear, is "an altered mother, carried away by her beastly influences." The infanticides are attributed to individual failings, even though the mostly likely explanation was that the mothers were responding to the conditions of captivity. Onelli and the unknown author of "Una pesquisa sensacional" did not question if motherhood is really a natural state or if poverty and captivity are malignant. Rather, some women just had "beastly influences" and could not be trusted. The future of the nation was in mothers' hands, but the animal cases showed those hands to be unsteady, further justifying control of women's minds, bodies, and motherhood by male scientists and doctors.

INTERLUDE

Zoo elephants seemed to be ideal creatures for codifying heterosexuality as national culture. In the first century CE, Pliny the Elder claimed the pachyderms were so modest that they "never mate except in secret" and that they only mated five times per year. Adultery and fighting over females were nonexistent.[1] Various medieval bestiaries continued this tradition, describing elephants as chaste, monogamous, and having sex only to bear children. At the Buenos Aires Zoo, the first elephants were a pair named Neán and Sayán, with Neán replaced by Nayán after she died. Nayán and Sayán successfully mated, engendering Phúa Victoria Porteña, symbol of the city's progress. In the zoo's journal and popular press, the adults were depicted as ideal parents behaving in gender appropriate ways, embodying the intimacy of heterosexual relations for the nation. Despite the ubiquity of these chaste stories, the 1904 Zoo Guide admitted that "our pair celebrated their wedding ceremony in the presence of five thousand spectators."[2] Given an elephant's gestational period of eighteen to twenty-two months, it is possible that those five thousand visitors saw the conception of Phúa Victoria, ultimate sign of the nation's future. What did it mean to violate the intimacy of heterosexual culture and have elephant sex in public?

As seen in the last chapter, familial animal groups in enclosures, the layout of park spaces, and zoo stories of marriage and reproduction attempted to codify reproductive heterosexuality as Argentine national culture. But the "biological exuberance" of the animal world, to use Bruce Bagemihl's term, and the fact that enclosures were designed to maximize humans' ability to see specimens (i.e., to deny them privacy) queered zoo spaces, putting sex and relationships of all sorts on display and chipping away at the presumed naturalness of national heterosexuality.[3]

One of the myriad purposes of zoos is to make animals available for the human gaze. In early zoos, including in Buenos Aires, this meant enclosures with few places for animals to hide. Concrete was a common building material, as were metalwork and iron bars. The Buenos Aires lion cage, for example, had a ground level made of cement with barred walls for people to peer through. The only other space the lions had access to were cramped underground cells. Often, there were also doors between an animal's indoor and outdoor spaces, allowing a zookeeper to force the animal to stay in view. Even in more open enclosures such as the guanaco corrals, the poor nature of the soil meant Holmberg and Onelli had great difficulty getting plants and trees to grow, limiting the number of natural hiding spots.

This reality meant that zoo visitors most certainly observed animal copulation, even though zoo stories avoided mentioning the actual sex act or cloaked it in euphemism when endlessly discussing animal romance and births. Onelli's elephants are the only explicit reference I have found in the archive to visitors seeing animal sex, but I am certain there were many more. The sheer number of births celebrated in the Buenos Aires Zoo between 1875 and 1924 tells of abundant relations. Some animals are prolific, with lions mating up to one hundred times a day during breeding season and goshawks mating "500–700 times for each clutch of eggs they produce."[4] And animals, like humans, have sex for nonbreeding purposes.[5] Indeed, even in today's more naturalistic exhibits with numerous places to hide, rare is the zoo visitor who has never observed animals having sex: one mounting another, the small child who innocently offers up "look, they're playing," and parents' stuttering attempts to respond. Although it was commonplace, zoo sex was taboo and thus buried in euphemism, elided completely, and/or hidden behind the privacy screen of legitimate marriage and the family unit.

As Berlant and Warner articulate, sex in public creates queer spaces; unsurprisingly, homosexuality is further absent in zoo stories. While there is debate over labels and definitions, scientists have documented that "the world is, indeed, teeming with homosexual, bisexual, and transgendered creatures of every stripe and feather."[6] And like zoo animals' misery, people have been aware of these realities for centuries, if not longer. Georges-Louis Leclerc, Comte de Buffon, was the first to make Western scientific observations of animal homosexuality in the eighteenth century. The nineteenth and early twentieth centuries gave rise to further studies that identified yet other examples, especially in birds and mammals.[7] Scientists have posited many explanations for homosexual and homosocial behavior in animals. These include controlling population size, dominance expression, social tension regulation, reconciliation, social bonding, alliance formation, acquisition of alloparental care, mate attraction, inhibition of competitors' reproduction, practice for heterosexual activities, kin selection, a functionless by-product of adaptations, maladaptation, or biological exuberance.[8]

Holmberg, Onelli, and their collaborators surely knew of the possibility of animal homosexual relations and must have observed them themselves, but they are even rarer than public sex in the zoo archive. Indeed, the only oblique reference I found does not refer to actual homosexual relations but instead uses them as an analogy for a heterosexual practice in tick species. In a speech on the parasitic arachnids, Onelli's frequent collaborator Fernando Lahille explained that "in the world of ticks, the fertilizing gesture cannot be called copulation, although in certain cases,—I think insufficiently elucidated, the male's face intervenes, at least in the preparation—Oh Lesbos!—if not in the actual act, as thought by De Greer (1775), Muller (1816), and Gené (1848)."[9] In referring to Lesbos, Lahille considers cunnilingus a lesbian practice, presumably assuming one would only participate in such a thing if penetration were not an option. "Homosexual behavior," as he understands it, is relegated to the lowest of low creatures, literal blood suckers.

Why was homosexual sex so completely erased from interpretations of the Buenos Aires Zoo, its animals, and the natural world? Of course, there was much less awareness of homosexual animal relations at the time. Additionally, however, homosexual relations were less useful for the various social projects and future projections carried out at the zoo. Building on Elizabeth Freeman's work in *Time Binds*, Szcygielska argues that the zoological display normalizes anthropocentric ideas of heterosexual family structures and relations in a future-oriented paradigm. Same-sex relationships, which would not end in a child, were thus "an evolutionary dead end."[10] In a country like Argentina where reproduction was seen as a civic duty, among other religious and social reasons for avoiding the topic, animals who practiced homosexual behaviors were failing to fulfill the duties of citizenship. They were not practical as models for constructing the nation.

While public copulation and homosexual animal behavior were both avoided, there was one fundamental difference. Most visitors likely could not differentiate between male and female specimens of many species. Many zoo employees were also similarly uninformed, for positioning in sexual activities does not always reveal an animal's biological sex.[11] Their gendering of zoo animals was almost entirely dependent on signage and their own assumptions. Visitors could thus continue to assume animals were heterosexual and would not question the nuclear family basis for their partnerships. The belief that homosexual behaviors were nonexistent in the animal world, and thus not natural, could continue to flourish. Sex was harder to avoid, and zookeepers and writers must have had to work to shape interpretations of animal sex in public for the human audience, requiring active intervention instead of passive avoidance.

CONCLUSION

A CENTER FOR THE IRRADIATION OF CULTURE AND IDEAS

The Buenos Aires Zoo as Municipal, National, and International Institution

Onelli's death in 1924 was the final mark on a slow slide from the zoo's peak. As the municipality invested in more and more *paseos*, the visitors that constituted the zoo's life blood diminished rapidly. Now the Palermo institution had to compete with proliferating movie theaters, skating rinks, and amusement parks.[1] Onelli himself turned his attention from the traditional aspects of his zoo, dedicating more and more time to developing the Escuela de Avicultura Práctica (School of Practical Aviculture), Escuela Municipal de Telares (Municipal Weaving School), and on a quixotic search for a plesiosaur in Patagonia. He also ceased publishing the *Revista del Jardín Zoológico de Buenos Aires* in 1922.

Onelli died on October 20, 1924, and his successor, Adolfo "Dago" Holmberg (grandnephew of Eduardo L. Holmberg), ordered the closure of the zoo from December 1924 to March 1, 1925, for urgent reforms.[2] Despite Holmberg's desire to give the animals more freedom, the public was unsatisfied with the pace of change. In 1930, Luis Góngora griped in *Mundo Argentino*, "as the world has turned, our Zoo, admirable thirty years ago, to the point of being taken into account, is regrettably backward today." It paled in comparison to the Sydney Zoo, itself in a "new city" like Buenos Aires. He concluded, "the reform of our Zoological Garden would be extremely useful."[3] In the same magazine the following year, David Rolla asked, "When will our zoo be modernized?" The article's subtitle was biting: "Animals displayed in small and unhygienic cages are subjected to pointless torture." He described sad animals and stereotypic behaviors. Both Góngora and Rolla used the word "neurasthenic" to describe the biggest beasts, marking their behaviors as pathological. Like Góngora, Rolla pointed to a foreign zoo as a model, although he chose Carl Hagenbeck's Tierpark in Stellingen (opened in 1907). For Rolla, the only correct way to display animals was in more naturalistic exhibits with the bars and cages removed.[4] The situation at the Buenos Aires Zoo became so bad that the mayor designated Dr. Silvio E. Parodi and Manuel Hortal Muñagorri to conduct a thorough investigation. Although Parodi exonerated Holmberg of any wrongdoing, it is evident just how far the Jardín Zoológico de Buenos Aires's star had fallen.[5]

Throughout this book, I have traced some of the many attempts to address social problems in late nineteenth- and early twentieth-century Buenos Aires through the zoo. Hygiene and public health were among the very first, codified in the institution's founding documents and early regulations. In the face of the chaos of the modern city, it would provide a clean, green space for residents to recover physically and mentally and prepare children to be strong future citizens and workers. The zoo's employees modeled hygiene, while the guidebooks provided information on vaccinations and infant feeding. Directors Holmberg and Onelli collaborated extensively with medical, veterinary, and carceral institutions in the city, demonstrating the importance of animal bodies for improving human health. Hygiene could also be weaponized for political fights: it was an important tool for justifying control over the lower classes or specific groups of immigrants. It could also be used to push for animal rights, as seen by Ignacio Albarracín's fight with Onelli.

Next, the zoo and the surrounding Parque Tres de Febrero were founded with egalitarian intentions. They were meant to be spaces for people of all social classes and ethnicities. But simply opening space did not erase conflict, and wealthier zoo visitors often complained about poorer visitors while using their presence at the zoo as a sign of the relative well-being of the immigrant and worker. At the same time, the popular trope of the aristocratic zoo animal was pressed into service to critique human and economic living conditions in the city. A wide range of visitors felt that zoo animals lived in nicer homes than people, ate higher quality food, and were waited on hand and foot despite not working. The socialist paper *La Vanguardia* also juxtaposed this treatment and the public's deep concern for the animals with the perceived lack of care for workers and their children. In contrast, Onelli used natural laws and animal behavior to trace a division between a legitimate socialism and a too radical, extreme form. The paternalist protections he elaborated for his animals and employees fell under the former category, while strike behavior and big changes to social structures occupied the latter and could not be supported by his observations of the natural world.

Many *porteño* workers were also immigrants, as was Onelli himself. As nineteenth-century promotion of immigration turned to twentieth-century reaping the effects of mass immigration, the zoo once again played a role. Scientific considerations of collection composition and the advisability of acclimatization were about animals, of course, but they also stood in for debates over the relative value of humans of varying origins. The zoo demonstrated that both immigrants and Creoles had a role in Argentina but insisted that cosmopolitanism should never be pursued at the expense of the "Argentine," an identity that was not aligned with Indigenous people even when expressed through analogy with autochthonous animals. The focus on the Creole bled into an articulation of a *raza latina*, which Onelli increasingly advocated for as a source of national identity. Through speeches, publications, and zoo architecture and art, he postulated an Argentina united by a cultural Latin identity that could, and indeed must, stand up to US economic interests and Protestant cultural invasion.

Finally, a generalized maternalization of women in Argentine society meant increased concern over their role in physical and social reproduction, both as an issue itself and in conjunction with hygiene, labor, and immigration. Children were the future, and mothers were the ones who would most profoundly shape them in the uterine and extrauterine environments. The display of zoo animals in family groupings and zoo stories depicting them as legitimately married fathers and mothers and satisfying gendered roles contributed to the solidification of a heterosexual national culture that posited that sex was for reproduction and that women were innately mothers. Zoo employees also modeled these behaviors, especially through care work. As with formulations regarding class conflict, many of these theses were based on observations of animal behavior that naturalized what were actually historically specific cultural forms. Animals that broke from the expected pattern, such as the rhea, were not interpreted as evidence of other gender expressions, but rather aberrations useful for denigrating more forceful strands of feminism.

Each of these themes can be traced throughout the period from at least 1888 to 1924, but the attention given to each changed in proportion and intensity over time. Hygiene was the most constant and present concern, appearing in many reports and stories of the zoo, as well as taking up a good chunk of the zoo's budget and personnel. It was a particular focus in the first thirty years of the zoo, likely reflecting city preoccupations and the real conditions of the zoo grounds. Built from essentially nothing, infrastructure like drainage, running water, and clear paths were sorely lacking for many years. Add to that the constant hygienic challenges of animal collections and massive numbers of visitors, and it is no surprise there was such a focus on hygiene. Debates over native and exotic species dated to the zoo's earliest years, but Onelli's personal defenses of the Creole and the *raza latina* grew much stronger over time. This may reflect changing perceptions of immigration, the rise of cultural nationalism, and/or an increasing general conservatism as Onelli grew older. He also grew more strident regarding communism and other labor movements. Interestingly, several of Onelli's texts on anti-imperialism and anti-communism were among his most reprinted and circulated in the decade after his death. Finally, the representations of gender at the zoo were relatively consistent over time, but the focus did shift from representations of animals to satellite sites and programs, such as the Escuela de Telares that aimed to preserve Creole culture, provide poor women with work that would allow them to contribute to the family economy while caring for children, and keep them away from unhygienic factories.

Three aspects remained relatively constant over the period and across topics. First, a sustained interest by zoo administrators, visitors, and journalists and writers in using the zoo and its animals to comment on human social conditions and effect change. Second, that the conditions of the zoo, malleability of animals for interpretation, and number of modalities of influence meant that the zoo truly could reach just about every type of person in the city. It was designed for adults and children, relatively cheap to get to and enter, and required no literacy nor mastery of Spanish to enjoy. Animals were appealing and fun, and the ultimate unknowability of other species meant they could become metaphors for competing images. The zoo also offered many pathways to shape opinions and behavior, including the structure of the physical grounds, animal displays and signage, the zoo guidebooks, programming, texts for popular and technical audiences, and numerous speeches by the directors, particularly Clemente Onelli, who reached the status of minor celebrity. Finally, throughout the period there is a clear tension between a desire to improve society and a fear of too much change. Just as Joseph Pierce writes of the cultural nationalism of Ramos Mejía, Ingenieros, and Bunge, the social project of the zoo participated in a "double discourse" where educating the masses was necessary for improving culture and institutions, but the transformational implications of an educated populace were terrifying.[6] There was no upsetting the status quo. On one hand, this reflects who ran the zoo: Holmberg and Onelli were both connected to the ruling elite and had to have municipal support for every single expense or project. Radical proposals would probably have seen them quickly removed from their posts. On the other hand, the resistance to change also reflects the structure of the institution itself. Zoos are built on hierarchies, particularly a human-animal ranking further subdivided by species in terms of their interest and draw, as well as power relations of the gaze. They also, as we have seen, serve to express other hierarchies such as class, race, and gender. As the case of the Buenos Aires Zoo shows, these institutional characteristics put limits on the reforms that could originate and be developed on the grounds.

At the same time, the top-down structure did not stop people and animals from trying to use the zoo on their own terms. Some appropriated materials, like when an angry former employee stole four boa constrictors to sell to the Teatro Chinesco.[7] The zoo was supposed to exercise a moralizing function, but reports of robberies and signs warning visitors of thieves suggest that some *porteños* took advantage of zoogoers' distraction for their personal benefit.[8] Other humans picked flowers and attacked animals. Animals also found their own ways to resist, including rejecting food, refusing to entertain the public, opening cages, escaping, biting, and even dying. These actions forced changes in employee behavior, the structure of the zoo, and the premature end or restriction of projects. For example, Onelli believed that walking on lawns would be the most hygienic for visitors. Nonetheless, he and his guards fought a continual battle to keep people on the pathways so their footsteps would not kill the grass. He also had to replace most of the flowers with trees and shrubs because visitors were pulling them out at the root.[9] Social projects at the zoo may have been ideated by those in power, but in the end were always negotiated as people and animals reacted and exerted their own power in whatever ways they could.

How do the trends I have analyzed at the Buenos Aires Zoo compare to its counterparts abroad? The interest in bettering society, promoting middle-class values, and balancing science and entertainment, as well as the problems with controlling animals and zoo visitors, span contexts. Holmberg and Onelli corresponded most with European zoo directors and traders and subscribed to European scientific journals, but in terms of its social functions, the Buenos Aires Zoo shows more similarities to US zoos. The zoo in Palermo was founded nearly contemporaneously with the Philadelphia and Cincinnati zoos, whereas many European zoos emerged decades earlier. Argentina and the US both have settler colonial backgrounds and long histories of fighting to control national territory and forcibly subdue native populations; Argentina even explicitly mimicked the US in designing and implementing Indigenous policies.[10] Mass immigration was also shaping both societies, particularly in large cities like Philadelphia, New York, and Buenos Aires. Consequently, both the immigrant animal trope and frankly racist claims that certain immigrant groups were problem visitors emerged in both.

There were also some significant differences. The Buenos Aires Zoo was a municipal zoo, but it stood for the nation in a way that did not happen with the first US zoos, likely because of different relationships between the capital cities and the interior and the fact that the US had many more large cities than Argentina. Class dynamics were also different, with distinct expressions of and reactions to labor agitation. Finally, the Buenos Aires Zoo was a space from which to articulate anti-imperial and particularly anti-US/anti-Anglo formulations in a way that was obviously not happening in the US itself. I am sure there are many more similarities and differences to both US and European zoos, and I look forward to learning about them from scholars of those regions.

The connections to other zoos in the Global South, particularly in Latin America, also remain to be seen. Onelli corresponded with zoo directors in Giza, Sydney, and Bombay, but the historical archive does not reveal much direct contact between Argentina and its peers. Perhaps most surprisingly, I have not found any correspondence with zoos in Brazil or Mexico, and very little with Argentina's neighbor Uruguay. This suggests that there were not strong institutional ties in the region, even though explorers and diplomats frequently trafficked animals across national borders. Truthfully, there has been so little published about Latin America zoos thus far that it is difficult to make comparisons. Luckily, there are many brilliant scholars working on these topics. I have no doubt that soon there will be rich texts that will undoubtably engage, support, and correct this one.

EXPANSION

One final note: throughout this book I have been fairly loose in my use of Buenos Aires and Argentina, slipping between the municipal and national levels when analyzing the representation and interpretation of "zoo-cial life." This usage echoes that of the zoo administrators and the journalists who wrote about the animals. For instance, Phúa Victoria Porteña was named for the city of her birth, but the national and international press highlighted her birth as an Argentine victory on the world stage. Additionally, foreign visitors to the zoo almost always interpreted what they saw there as evidence of Argentine society, not just that of Buenos Aires.

But the fact remains that Buenos Aires is not Argentina, as anyone from the interior will tell you. So did the *porteño* zoo interact with the provinces, and if so, how? As discussed in Chapter 3, the zoo received animals from other parts of the country, particularly those being actively colonized. The *Revista del Jardín Zoológico de Buenos Aires* also published studies by "provincial scholars who find in it a place to make themselves known." Onelli was proud of this fact and claimed that one of his goals was to "link the interior and the metropolis as much as possible through a community of studies, ideals, and culture."[11] In exchange, the journal was distributed to provincial scholarly institutions and the zoo guide was used as a textbook in the provinces of Buenos Aires, Santa Fe and Entre Ríos, sharing those depictions and projects with populations beyond the capital.[12] The zoo also sent animals to the provinces for various acclimatization projects, such as when Onelli worked with the governor of the Territorio Nacional de los Andes (National Territory of the Andes) to acclimatize chinchillas and karakul sheep.[13]

The Buenos Aires Zoo also expanded its national sphere of influence by collaborating with the development of zoos in other Argentine cities. Onelli cooperated with the founding, development, or provision of animals in Mendoza (1903), Bahía Blanca (1903), La Plata (1907), Córdoba (1915), and Tucumán (originally founded in 1905 but relocated and refounded in 1916).[14] A few years after the foundation of the zoo in La Plata, capital of Buenos Aires province, Onelli reflected proudly on his participation: "Our dreams were realized: we always wanted the Jardín Zoológico de Buenos Aires to be a center for the irradiation of culture and ideas, to break with the mold of localist egoism—the so-called *patria chica*—to serve in its modest sphere the good ideals of the *patria grande*."[15] This collaboration continued with the exchange of animals, sometimes permanently and sometimes temporarily, such as when the La Plata zoo sent a tigress to Buenos Aires to be bred with the capital's male.[16]

Onelli spent a good chunk of 1915 traveling to Córdoba and Tucumán to help establish or reformulate zoos in those cities. In Tucumán, the director of public works and director of parks had drawn up plans, but the governor, Dr. Ernesto Padilla, requested Buenos Aires's help with the practical details. Onelli gladly collaborated on the zoo and a model farm in Parque Centenario, which was designed to balance the climatic and cultural realities of Tucumán with "all the European perfections and all the most rigorous prophylactic hygiene measures."[17] The Tucumán park was a perfect example of Onelli's insistence on remaining modern without ever losing sight of the value of the *criollo*. He even sent a "family of exemplary workers" from Buenos Aires to take care of the farm and be role models.[18]

Dr. Ramón J. Cárcano, governor of Córdoba, was also interested in bringing a zoo to his city. In a letter to Onelli, he was insistent on the benefit of such an institution and his desire that it be done well so that it would serve the population long after his term in government ended. To that effect, he asked Onelli to visit, gather his impressions of the planned location and city, and give a speech to drum up public favor.[19] Onelli did as asked, delivering a speech, "Animals in Public and Private Life," at the Rivera Indarte theater on July 18, 1915. He argued that the "cordobesitos" were just as deserving of instruction and a zoo as the "porteñitos," as were the tired and overworked adults of that city.[20] Córdoba's ravines would provide a perfect natural location for the institution, while the city's long academic history meant there were many scholars who could take advantage of the zoo's scientific riches.[21] Governor Cárcano had ten thousand copies of Onelli's speech printed. The propaganda was successful and Cárcano presided over a December 25, 1915, inauguration ceremony attended by thirty thousand people.[22]

In later years, Onelli also supplied plans, advice, and animals to zoological projects in Bahía Blanca and Mendoza, again with the goal of encouraging the spread of the "natural influence of the capital to the territory of the entire Republic."[23] In these requests and in Onelli's words, we see an exportation of *porteño* culture and ideas, including hygiene, the relative value of the native and the exotic, and the benefit of moral and intellectual education in zoology, as well as interest in helping (and controlling) the working classes. Future projects will have to uncover how these provincial zoos mimicked or departed from the Buenos Aires experience, as well as the intersections between zoos and local conditions. It will also be beneficial to elaborate the networks between Buenos Aires and the provinces, as well as between provincial zoos themselves, to develop a complete understanding of the roles Argentine zoos and their animals played in influencing society from their foundations to the present.

NOTES

INTRODUCTION

1. Clemente Onelli, "Sed compasivo con los animales," *Revista del Jardín Zoológico de Buenos Aires: Época II* 7, no. 28 (December 1911): 348. The journal had not been published for a few years and Onelli restarted it in 1905; *Época II* (second era/series) distinguishes this run from the original one in the 1890s. All translations are mine unless otherwise noted. When the Spanish word choice is particularly relevant to the meaning or difficult to translate, I also include the quotation in its original language.
2. Clemente Onelli, "Una mirada retrospectiva: El Jardín Zoológico en el 1905," *Revista del Jardín Zoológico de Buenos Aires: Época II* 1, no. 4 (December 1905): 401.
3. Lisa Uddin, *Zoo Renewal: White Flight and the Animal Ghetto* (Minneapolis: University of Minnesota Press, 2015), 8.
4. J. M. Cinaghi, "Cuadros estadísticos comparativos," *Revista del Jardín Zoológico de Buenos Aires. Época II* 9, no. 34–36 (December 1913): 227.
5. See, for example, María Silvia Di Liscia and Ernesto Lázaro Bohoslavsky, *Instituciones y formas de control social en América Latina, 1840–1940: Una revisión* (Buenos Aires: Prometeo, 2005), 9–10.
6. Éric Baratay and Elisabeth Hardouin-Fugier's *Zoo: A History of Zoological Gardens in the West* documents this history in an encyclopedic format. In it, "the West" is narrowly defined as Europe with some inclusion of the United States. Baratay and Hardouin-Fugier, *Zoo: A History of Zoological Gardens in the West* (London: Reaktion Books, 2002). The second edition of Vernon N. Kisling's *Zoo and Aquarium History: Ancient Animal Collections to Conservation Centers* is more comprehensive, containing sections on Eastern Europe/Russia, the Middle East, Israel, Asia, India, China, Japan, Africa, and Latin America. Nonetheless, whole regions such as "Africa" and "Latin America" are given as much space as Great Britain and Ireland or the United States. Vernon N. Kisling, ed., *Zoo and Aquarium History: Ancient Animal Collections to Conservation Centers*, 2nd ed. (New York: CRC Press, 2022).
7. Adrián Gorelik, *The Grid and the Park: Public Space and Urban Culture in Buenos Aires, 1887–1936*, trans. Natalia Majluf (Pittsburgh, PA: Latin America Research Commons, 2022), 22–24.
8. Antonio Carbone, *Park, Tenement, Slaughterhouse: Elite Imaginaries of Buenos Aires, 1852–1880* (Frankfurt: Campus Verlag, 2022), 165.

9. In Sergio Pedernera and Estela Pagani, eds., *Sarmiento, espacio y política: El Parque 3 de Febrero* (Buenos Aires: Dirección General Patrimonio e Instituto Histórico de la Ciudad de Buenos Aires, 2010), 26. The accompanying CD includes images of the original document in Sarmiento's hand.
10. Eduardo Esparrach and Juan Carlos Sassaroli, *Historia del Zoológico de Buenos Aires: 1888–2020* (Buenos Aires: Vázquez Mazzini Editores, 2022), 30.
11. For more on Holmberg's position in the Argentine sciences, see Paula Bruno, "Eduardo L. Holmberg en la escena científica argentina: Ideas y acciones entre la década de 1870 y el fin-de-siglo," *Saber y Tiempo* 1, no. 1 (2015): 118–40.
12. For a history of the zoo's structures, see Marcela Liliana Díaz and Cristina Fernández, *Guía del patrimonio cultural del Jardín Zoológico de la Ciudad de Buenos Aires* (Buenos Aires: DyPContenidos, 2016).
13. Esparrach and Sassaroli, *Historia del Zoológico*, 31.
14. My calculation, based on Alejandro Pavlovsky, "Sección administrativa," *Revista del Jardín Zoológico de Buenos Aires: Época II* 1, no. 3 (October 1905): 313; and "Cuadro demonstrativo de la venta de entradas al Jardín Zoológico durante los últimos nueve años (1901–1909)," *Revista del Jardín Zoológico de Buenos Aires: Época II* 6, no. 20 (January 1910): 382.
15. Eduardo L. Holmberg, "Informe anual del Director del Jardín Zoológico correspondiente al año 1893," *Revista del Jardín Zoológico de Buenos Aires* 2, no. 1 (January 15, 1894): 28.
16. "Jardín Zoológico," *Caras y Caretas* 7, no. 280 (February 13, 1904).
17. Esparrach and Sassaroli, *Historia del Zoológico*, 52.
18. Diego del Pino provides a comprehensive biography of Onelli in *Clemente Onelli: El más criollo de los tanos* (Buenos Aires: Ediciones Turísticas, 2005).
19. Aldo Mario Giudice, Eduardo Esparrach, and María Teresa Ferrero de Roqué, "Serie: Elefantes rioplatenses y zoológicos: La inecuación perfecta. (III) Elefantes en el Zoológico de Buenos Aires," *Revista Boletín Biológica*, no. 39, año 12 (2018): 12. On Mimí the giraffe, see "La girafa 'Mimí,'" *Caras y Caretas* 15, no. 709 (May 4, 1912).
20. Cinaghi, "Cuadros estadísticos comparativos," 227.
21. "Cuadro demonstrativo de la venta de entradas al Jardín Zoológico desde el año 1914 a 1921," *Revista del Jardín Zoológico de Buenos Aires: Época II* 18, no. 66–68 (October 1922): 246. I have not found data for 1922 and beyond, but there is no reason to think that the number of visitors to the zoo would have diminished dramatically in the last years of Onelli's life.
22. Oliver Hochadel, "A Global Player from the South: The Jardín Zoológico de Buenos Aires and the Transnational Network of Zoos in the Early Twentieth Century," *História, Ciências, Saúde-Manguinhos* 29, no. 3 (2002): 789–812.
23. Notable visitors included Spanish writers Vicente Blasco Ibáñez and Santiago Rusiñol, former French prime minister Georges Clemenceau, Uruguayan author Horacio Quiroga, and Infanta Isabel, Countess of Girgenti. The first four published accounts of their visits. Vicente Blasco Ibáñez, *Argentina y sus grandezas* (Madrid: Editorial Española-Americana, 1910); Santiago Rusiñol, *Un Viaje al Plata*, trans. G. Martínez Sierra (Madrid: V. Prieto, 1911); Georges Clemenceau, *Notas de viaje por la América del Sur: Argentina, Uruguay, Brasil*, trans. Miguel Ruiz (Buenos Aires: Cabut, 1911). On Quiroga's writing about the zoo for the magazine *Caras y Caretas*, see Alejandro C. Eujanian, *Historia de revistas argentinas, 1900–1950: La conquista del pueblo* (Buenos Aires: Asociación Argentina de Editores de Revistas, 1999), 44. Quiroga's short story "La tortuga gigante," found in *Cuentos de la selva* (1918), involves a tortoise who rescues an ill man, carry-

ing him to Buenos Aires on his back. The tortoise then lives happily ever after in the zoo, cared for by the man's good friend, the zoo director. Horacio Quiroga, "La tortuga gigante," in *Cuentos de la selva* (Buenos Aires: Buenos Aires Cooperativa Editorial Limitada, 1918), 5–15.

24. Benjamin Bryce, "Paternal Communities: Social Welfare and Immigration in Argentina, 1880–1930," *Journal of Social History* 49, no. 1 (2015): 216–17.
25. For an overview of the many changes in the city see José Luis Romero and Luis Alberto Romero, eds., *Buenos Aires: Historia de cuatro siglos*, vol. 2 (Buenos Aires: Editorial Abril, 1983).
26. Patricia Méndez and Rodrigo Gutiérrez Viñuales, "Buenos Aires en el Centenario: Edificación de la nación y la nación edificada," *Apuntes* 19, no. 2 (2007): 221–27.
27. Juan Suriano, ed., *La cuestión social en Argentina, 1870–1943* (Buenos Aires: Editorial La Colmena, 2000), 2.
28. Carl Solberg, "Immigration and Urban Social Problems in Argentina and Chile, 1890–1914," *Hispanic American Historical Review* 49, no. 2 (May 1969): 215.
29. Julia Rodríguez, *Civilizing Argentina: Science, Medicine, and the Modern State* (Chapel Hill: University of North Carolina Press, 2006), 7.
30. Mario Zavattaro, "Los hombres prácticos," *Caras y Caretas* 11, no. 532 (December 12, 1908).
31. Clemente Onelli, "Variantes sobre nuestro Jardín Zoológico," *Revista del Jardín Zoológico de Buenos Aires: Época II* 8, no. 30 (July 1912): 176.
32. "En el Zoológico," *PBT* 10, no. 440 (May 3, 1913).
33. Clemente Onelli, "El Jardín Zoológico en 1916," *Revista del Jardín Zoológico de Buenos Aires: Época II* 12, no. 48 (December 1916): 507.
34. Nora Coiticher, "Picnics y fiestas campestres: Espacios de ocio y de derroche en la ciudad anarquista (Buenos Aires, 1897–1910)" Instituto de Arte Americano e Investigaciones Estéticas, Seminario de Crítica, no. 205, Buenos Aires, May 13, 2016.
35. Eduardo L. Holmberg, "El reglamento del Jardín Zoológico," *Revista del Jardín Zoológico de Buenos Aires* 1, no. 1 (January 15, 1893): 18.
36. Clemente Onelli, "El Jardín Zoológico en 1907: Vistas para el futuro," *Revista del Jardín Zoológico de Buenos Aires: Época II* 3, no. 12 (December 1907): 364.
37. Baratay and Hardouin-Fugier, *Zoo*, 99.
38. Clemente Onelli, "La sucursal sur del Jardín Zoológico: Su objeto," *Revista del Jardín Zoológico de Buenos Aires: Época II* 3, no. 11 (October 1907): 271.
39. Randy Malamud, *Reading Zoos: Representations of Animals and Captivity* (New York: New York University Press, 1998), 5.
40. Eujanian, *Historia de revistas,* 21–23.
41. Takashi Ito, "History of the Zoo," in *Handbook of Historical Animal Studies*, ed. Mieke Roscher, André Krebber, and Brett Mizelle (Berlin: De Gruyter Oldenbourg, 2021), 450.
42. Amalia Cardona Leites, "El fenómeno editorial *Caras y Caretas* y el papel del escritor en sus páginas: Horacio Quiroga y S. Fragoso Lima," *Questión: Revista Especializada en Periodismo y Comunicación* 1, no. 48 (December 2015): 81–82.
43. Quoted in Eujanian, *Historia de revistas*, 44.
44. Clemente Onelli to Señor Secretario, March 13, 1914, Copiador 1913–1914, 343, Ecoparque, Buenos Aires.
45. Erica Fudge, *Brutal Reasoning: Animals, Rationality, and Humanity in Early Modern England* (Ithaca, NY: Cornell University Press, 2006), 190.
46. Gabriel Giorgi, *Formas comunes: Animalidad, cultura, biopolítica* (Buenos Aires: Eterna Cadencia, 2014), 15.

47. Garry Marvin, "Wolves in Sheep's (and Others') Clothing," in *Beastly Natures: Animals, Humans, and the Study of History*, ed. Dorothee Brantz (Charlottesville: University of Virginia Press, 2010), 62.
48. Cary Wolfe, *What Is Posthumanism?* (Minneapolis: University of Minnesota Press, 2009), xv–xvi.
49. Éric Baratay, *Animal Biographies: Toward a History of Individuals*, trans. Lindsay Turner (Athens: University of Georgia Press, 2022).
50. Martha Few and Zeb Tortorici, eds., *Centering Animals in Latin American History* (Durham, NC: Duke University Press, 2013), 19.
51. See, for example, J. Barrington-Johnson, *The Zoo: The Story of London Zoo* (London: Hale, 2005) and Catherine de Courcy, *Dublin Zoo: An Illustrated History* (Cork: Collins, 2009).
52. Oliver Hochadel, "Science at the Zoo: An Introduction," *Centaurus* 64, no. 3 (October 2022): 561–90, contains an excellent bibliography of scientific endeavors in worldwide zoos.
53. Juliana Adelman, *Civilised by Beasts: Animals and Urban Change in Nineteenth-Century Dublin* (Manchester: Manchester University Press, 2020); Daniel E. Bender, *The Animal Game: Searching for Wildness at the American Zoo* (Cambridge, MA: Harvard University Press, 2016); Gary Bruce, *Through the Lion Gate: A History of the Berlin Zoo* (New York: Oxford University Press, 2017); Andrew Flack, *The Wild Within: Histories of a Landmark British Zoo* (Charlottesville: University of Virginia Press, 2018); David Grazian, *American Zoo: A Sociological Safari* (Princeton, NJ: Princeton University Press, 2015); Elizabeth Hanson, *Animal Attractions: Nature on Display in American Zoos* (Princeton, NJ: Princeton University Press, 2002); Takashi Ito, *London Zoo and the Victorians, 1828–1859* (Woodbridge, Suffolk: Royal Historical Society, 2014); Ian Jared Miller, *The Nature of the Beasts: Empire and Exhibition at the Tokyo Imperial Zoo* (Berkeley: University of California Press, 2013); Bob Mullan and Garry Marvin, *Zoo Culture*, 2nd ed (Urbana: University of Illinois Press, 1999); Nigel Rothfels, *Savages and Beasts: The Birth of the Modern Zoo* (Baltimore, MD: Johns Hopkins University Press, 2002); and Daniel Vandersommers, *Entangled Encounters at the National Zoo: Stories from the Animal Archive* (Lawrence: University Press of Kansas, 2024).
54. Maneesha Deckha, "Toward a Postcolonial, Posthumanist Feminist Theory: Centralizing Race and Culture in Feminist Work on Nonhuman Animals," *Hypatia* 27, no. 3 (Summer 2012): 540.
55. Tracy McDonald and Daniel Vandersommers, eds., *Zoo Studies: A New Humanities* (Montreal: McGill-Queen's University Press, 2019).
56. Regina Horta Duarte, "Zoos in Latin America," *Oxford Research Encyclopedia of Latin American History*, September 26, 2017.
57. Flack, *The Wild Within*, 8.
58. Hanson, *Animal Attractions*, 9.
59. Jason Hribal, *Fear of the Animal Planet: The Hidden History of Animal Resistance* (Edinburgh: AK, 2010), 26.
60. Few and Tortorici, *Centering Animals*, 14.
61. Fudge, *Brutal Reasoning*, 190.

CHAPTER 1

1. "La higiene en el Zoo," *Billiken* 3, no. 143 (August 14, 1922).
2. On Mexico, see Ana Cecilia Rodríguez de Romo and Martha Eugenia Rodríguez Pérez, "Historia de la salud publica en México: Siglos XIX y XX," *História, Ciências,*

Saúde-Manguinhos 5, no. 2 (October 1998): 293–310, and Claudia Agostoni, *Monuments of Progress: Modernization and Public Health in Mexico City, 1876–1910* (Calgary: University of Calgary Press, 2003). For Brazil, see Gilberto Hochman, *The Sanitation of Brazil: Nation, State, and Public Health, 1889–1930*, trans. Diane R. Grosklaus Whitty (Urbana: University of Illinois Press, 2016) and Okezi T. Otovo, *Progressive Mothers, Better Babies: Race, Public Health, and the State in Brazil, 1850–1945* (Austin: University of Texas Press, 2016). Many of the essays in Diego Armus's edited collection address hygiene across the Americas. Diego Armus, ed., *Entre médicos y curanderos: Cultura, historia y enfermedad en la América Latina moderna* (Buenos Aires: Grupo Editorial Norma, 2002).

3. Manuel Durán, "Sexualidad, producción y trabajo en el discurso higienista y eugenésico en Chile y Argentina, 1860–1930," *Revista Nomadías* 23 (July 2017): 37.
4. Rodríguez, *Civilizing Argentina*, 187.
5. Alejandro Kohl, *Higienismo argentino: Historia de una utopía. La salud en el imaginario colectivo de una época* (Buenos Aires: Editorial Dunken, 2006), 44.
6. Diego Armus, *The Ailing City: Health, Tuberculosis, and Culture in Buenos Aires, 1870–1950* (Durham, NC: Duke University Press, 2011), 232, 242. See also Eduardo Zimmerman, "Racial Ideas and Social Reform: Argentina, 1890–1916," *Hispanic American Historical Review* 72, no. 1 (1992): 23–46.
7. Armus, *The Ailing City*, 243.
8. Rodríguez, *Civilizing Argentina*, 178.
9. On hygiene and socialists, see Dora Barrancos, *La escena iluminada: Ciencias para trabajadores, 1890–1930* (Buenos Aires: Plus Ultra, 1996), 41. For oligarchs, see Kohl, *Higienismo argentino*, 55. For feminism and hygiene, see Asunción Lavrin, *Women, Feminism, and Social Change in Argentina, Chile, and Uruguay, 1890–1940* (Lincoln: University of Nebraska Press, 1995), 97. For educators, see María Silvia di Liscia and Graciela Nélida Salto, eds., *Higienismo, educación y discurso en la Argentina, 1870–1940* (Santa Rosa, La Pampa: Editorial de la Universidad Nacional de La Pampa, 2004).
10. Armus, *The Ailing City*, 115, 119.
11. Holmberg, "El reglamento del Jardín Zoológico," 6.
12. Kohl, *Higienismo argentino*, 73.
13. Holmberg, "Informe anual . . . 1893," 3.
14. Clemente Onelli, "Vista retrospectiva: El Jardín Zoológico en 1920," *Revista del Jardín Zoológico de Buenos Aires: Época II* 17, no. 65 (April 1921): 112.
15. Richard J. Walter, *Politics and Urban Growth in Buenos Aires, 1910–1942* (Cambridge: Cambridge University Press, 1993), 11.
16. Ronaldo Munck, *Argentina: From Anarchism to Peronism* (London: Zed Books, 1987), 30–31.
17. Leandro H. Gutiérrez and Juan Suriano, "Workers' Housing and Living Conditions in Buenos Aires, 1880–1930," in *Essays in Argentine Labor History, 1870–1930*, ed. Jeremy Adelman (Basingstoke: Macmillan, 1992), 38–39.
18. Gutiérrez and Suriano, "Workers' Housing," 40.
19. Santiago Estrada, *Viajes y otras páginas literarias*, 4th ed., ed. Ricardo Ryan (Buenos Aires: Ángel Estrada, 1942), 115.
20. Eugenia Scarzanella, *Ni gringos ni indios: Inmigración, criminalidad y racismo en la Argentina, 1890–1940*, trans. Irene Theiner (Bernal, Buenos Aires: Universidad Nacional de Quilmes Ediciones, 2003), 91.
21. Armus, *The Ailing City*, 10.

22. Agapito Candileja, "Interview a S.M. Cónsul III," *Caras y Caretas* 9, no. 402 (June 16, 1906).
23. Gorelik, *The Grid and the Park*, 47.
24. Clemente Onelli to Señor Secretario de la Intendencia Municipal, October 28, 1916, Copiador 1916–1917, 141–142, Ecoparque, Buenos Aires.
25. Eduardo Marthur, "Pic-nic popular," *Caras y Caretas* 17, no. 811 (April 18, 1914).
26. "Primera fiesta del Animal," *El Monitor de Educación Común* 28, no. 425 (May 31, 1908): 294.
27. "Primera fiesta del Animal," 292.
28. Quoted in Marta Alicia Libedinsky, "La excursión escolar en la Ciudad de Buenos Aires (fines del siglo XIX y segunda década del siglo XX): Diálogos histórico-didácticos entre pasado, presente y viceversa" (PhD diss., Tandil, Universidad Nacional del Centro de la Provincia de Buenos Aires, 2021), 156.
29. Clemente Onelli to Señor Secretario de la Intendencia Municipal, March 4, 1918, Copiador 1917–1918, 432–34, Ecoparque, Buenos Aires.
30. Clemente Onelli to Señor Administrador General de Suministros, May 28, 1919, Copiador 1918, 1918–1919, 703, Ecoparque, Buenos Aires.
31. Clemente Onelli, February 25, 1919, Copiador 1918, 1918–1919, 430, Ecoparque, Buenos Aires.
32. *Guía popular del Jardín Zoológico Municipal de Buenos Aires 1904* (Buenos Aires: A. Rothkopf, 1904), 16.
33. *Guía popular . . . 1904*, 17.
34. Clemente Onelli, "Una mirada retrospectiva: El Zoo en 1906," *Revista del Jardín Zoológico de Buenos Aires: Época II* 2, no. 8 (December 1906): 314.
35. *Guía popular . . . 1904*, 17.
36. Pablo Barlaro, "Tratado de higiene práctica elemental (1a parte)," *Revista del Jardín Zoológico de Buenos Aires: Época II* 13, no. 51 and 52 (December 1917): 305–93.
37. María Silvia Di Liscia, "Marcados en la piel: Vacunación y viruela en Argentina (1870–1910)," *Ciencia & Saúde Coletiva* 16, no. 2 (2011): 415.
38. Di Liscia, "Marcados en la piel," 411.
39. Norma Isabel Sánchez, *La higiene y los higienistas en la Argentina: 1880–1943* (Buenos Aires: Sociedad Científica Argentina, 2007), 176.
40. *Guía oficial ilustrada del Jardín Zoológico Municipal de Buenos Aires 1908–1909* (Buenos Aires: Gaibisso, 1908), 3.
41. *Guía oficial . . . 1908–1909*, 49, 81.
42. *Guía oficial . . . 1908–1909*, 77.
43. *Guía oficial . . . 1908–1909*, 29, 5, 31.
44. *Guía oficial . . . 1908–1909*, 25, 47.
45. Armus, *The Ailing City*, 144.
46. *Guía oficial ilustrada del Jardín Zoológico Municipal de Buenos Aires 1920–1921* (Buenos Aires: Compañía General de Fósforos, 1920), 32.
47. Armus, *The Ailing City*, 130.
48. Clemente Onelli, "El Jardín Zoológico en el extranjero," *Revista del Jardín Zoológico de Buenos Aires: Época II* 5, no. 19 (October 1909): 270.
49. Clemente Onelli to Señor Gefe de la Oficina de Vacuna de la Asistencia Pública, October 26, 1921, Copiador 1920–1921, 879, Ecoparque, Buenos Aires.
50. Pavlovsky, "Sección administrativa," 314.
51. Clemente Onelli, "El Jardín Zoológico en el año 1908 y 1909," *Revista del Jardín Zoológico de Buenos Aires: Época II* 4, no. 16 (December 1908): 399.

52. Pedernera and Pagani, *Sarmiento*, 131; *Guía oficial . . . 1920–1921*, 141.
53. Díaz and Fernández, *Guía del patrimonio*, 114.
54. Onelli, "El Jardín Zoológico en el año 1908 y 1909," 399.
55. Manuel Gálvez, *Nacha Regules*, ed. Aníbal Jarkowski (Buenos Aires: Eterna Cadencia, 2010), 272.
56. Díaz and Fernández, *Guía del patrimonio*, 130.
57. Onelli, "El Jardín Zoológico en 1916," 509.
58. Clemente Onelli to Señor Secretario de la Intendencia Municipal, February 8, 1917, Copiador 1916–1917, 207, Ecoparque, Buenos Aires.
59. "Hablando con Onelli," *PBT* 14, no. 640 (March 3, 1917).
60. "Hablando con Onelli."
61. Barrancos, *La escena iluminada*, 18.
62. Fernando Lahille, "La zoología, los conocimientos que necesita y los estudios que abarca," *Revista del Jardín Zoológico de Buenos Aires: Época II* 3, no. 10 (July 1907): 127.
63. Clemente T. Andrade, F. E. Giacomotti, and D. I. Espinosa, "Excursión al Jardín Zoológico," *Monitor de Educación Común* 419, no. 25 (November 30, 1907): 341.
64. Adolfo Banchero, Coreolano Brea, and Río Raúl Franchi, "Excursión al Jardín Zoológico," *Monitor de Educación Común* 25 (1905): 250–57.
65. Andrade, Giacomotti, and Espinosa, "Excursión al Jardín Zoológico," 341.
66. Ángel P. Bonetti, *De la República Argentina y sus detractores: Homenaje a la República en su primer centenario de independencia* (Buenos Aires: El Imparcial, 1910), 163.
67. Barrancos, *La escena iluminada*, 39–40.
68. Barrancos, *La escena iluminada*, 51.
69. Alejandro Martín Yaverovski, "Tres repúblicas barriales y sus universidades: Las Universidades Populares en la ciudad de Buenos Aires en el período de entreguerras, un esquema preliminar," *Historia Caribe* 16, no. 38 (June 2021): 46.
70. "Primera fiesta del Animal," 300.
71. "Primera fiesta del Animal," 297.
72. "Primera fiesta del Animal," 297–98.
73. Onelli, "Sed compasivo," 348.
74. Sarmiento Sociedad Protectora de Animales, "29 de abril—Día del Animal," *Billiken* 4, no. 180 (April 30, 1923): 19.
75. "Celebración del Día del Animal," *Billiken* 4, no. 181 (May 7, 1923): 20. For more on the relationship between Billiken magazine, civics, and Argentine society, see Lauren Rea, *Children's Culture and Citizenship in Argentina: A History of Billiken Magazine (1919–2019)* (York: White Rose University Press, 2023).
76. "Comités Billiken," *Billiken* 6, no. 267 (December 29, 1924).
77. Zoos worldwide fretted about tuberculosis, particularly in primates. The disease killed valuable specimens and zoo employees, and scientists worried about zoonotic spread. Vandersommers, *Entangled Encounters*, 209.
78. Clemente Onelli, "Aumento vegetativo: Los principales productos del año," *Revista del Jardín Zoológico de Buenos Aires: Época II* 3, no. 9 (April 1907): 68.
79. Clemente Onelli, "El Jardín Zoológico y sus anexos en 1915," *Revista del Jardín Zoológico de Buenos Aires: Época II* 11, no. 44 (December 1915): 434.
80. William Evans of the Lincoln Park Zoo in Chicago was the first to try keeping monkeys outside during the winter, in 1904. It was not until the 1910s that other zoos followed his example and/or built open-air buildings. Vandersommers, *Entangled Encounters*, 227; Hochadel, "A Global Player," 790.

81. Eduardo L. Holmberg, "Neán: La elefante del Jardín Zoológico," *Revista del Jardín Zoológico de Buenos Aires* 1, no. 3 (March 15, 1893): 65.
82. *Guía popular . . . 1904*, 64.
83. Clemente Onelli, "La elefantita americana," *Revista del Jardín Zoológico de Buenos Aires. Época II* 2, no. 5 (April 1906): 20; Clemente Onelli, "Idiosincrasias individuales de los pensionistas del Jardín Zoológico," *Revista del Jardín Zoológico de Buenos Aires. Época II* 3, no. 11 (October 1907): 196.
84. Clemente Onelli, "Plagas agrícolas y paludismo," *Revista del Jardín Zoológico de Buenos Aires. Época II* 8, no. 29 (April 1912): 33.
85. Onelli, "Plagas agrícolas," 37.
86. Eduardo L. Holmberg, "Osteomalacia," *Revista del Jardín Zoológico de Buenos Aires* 1, no. 2 (February 15, 1893): 39.
87. Holmberg, "Neán," 72.
88. Clemente Onelli, "Apuntes sobre el Jardín Zoológico para el Dr. Gustavo Loisel del Laboratorio de Embriología de París," *Revista del Jardín Zoológico de Buenos Aires. Época II* 3, no. 11 (October 1907): 283.
89. Onelli, "El Jardín Zoológico y sus anexos," 431.
90. Clemente Onelli, "El Jardín Zoológico en 1909," *Revista del Jardín Zoológico de Buenos Aires. Época II* 6, no. 21 (April 1910): 49–50.
91. Clemente Onelli, "Sección médica," *Revista del Jardín Zoológico de Buenos Aires. Época II* 1, no. 1 (April 1905): 87.
92. Onelli, "El Jardín Zoológico en 1909," 50.
93. Onelli, "El Jardín Zoológico y sus anexos," 431–32.
94. Clemente Onelli to Señor Secretario de la Intendencia Municipal de la Capital, January 10, 1905, Copiador 1903–1904–1905, 416–17, Ecoparque, Buenos Aires.
95. Onelli first mentions coprology in the *Revista del Jardín Zoológico* in July 1906 in an article including several case studies of his work with Ricardo Lynch. Clemente Onelli, "Exámenes coprológicos en los animales," *Revista del Jardín Zoológico de Buenos Aires. Época II* 2, no. 6 (July 1906): 117–24.
96. Ricardo Lynch, "L'alimentation et l'instinct de l'alimentation chez l'homme et chez les animaux," *Revista del Jardín Zoológico de Buenos Aires. Época II* 7, no. 25 (April 1911): 15.
97. Clemente Onelli, "Utilidad de la coprología: Algunas autopsias que lo confirman," *Revista del Jardín Zoológico de Buenos Aires. Época II* 2, no. 8 (December 1906): 281.
98. Gustave Loisel, *Histoire des ménageries de l'antiquité à nos jours*, vol. 3 (Paris: Octave Doin et Fils, 1912).
99. Clemente Onelli, "Idiosincrasias individuales de los pensionistas del Jardín Zoológico," *Revista del Jardín Zoológico de Buenos Aires. Época II* 8, no. 33 (April 1913): 12.
100. Onelli, "El Jardín Zoológico en 1909," 50.
101. Christofredo Jakob and Clemente Onelli, *Atlas del cerebro de los mamíferos de la República Argentina: Estudios anatómicos, histológicos y biológicos comparados, sobre la evolución de los hemisferios y de la corteza cerebral* (Buenos Aires: Guillermo Kraft, 1913), 3.
102. Mariana Dovio, "El Instituto de Criminología y la 'mala vida' entre 1907 y 1913," *Anuario de la Escuela de Historia Virtual* 4, no. 4 (2013): 99.
103. Ricardo Donato Salvatore and Carlos Aguirre, eds., *The Birth of the Penitentiary in Latin America: Essays on Criminology, Prison Reform, and Social Control, 1830–1940* (Austin: University of Texas Press, Institute of Latin American Studies, 1996), 22.
104. Dovio, "El Instituto de Criminología," 103.

105. Christofredo Jakob, "Problemas actuales de psiquiatría general y sus relaciones con las ciencias sociales y jurídicas," *Revista de Criminología, Psiquiatría y Medicina Legal* 4 (1917): 13.
106. Clemente Onelli, "Foseta del vermis en los cráneos de indios, mamíferos y saurianos," *Revista del Jardín Zoológico de Buenos Aires. Época II* 1, no. 4 (December 1905): 353.
107. Onelli, "Foseta del vermis," 353.
108. Onelli, "Foseta del vermis," 356.
109. Clemente Onelli, "Correspondencia," *Revista del Jardín Zoológico de Buenos Aires. Época II* 4, no. 14 (July 1908): 202.
110. Cesare Lombroso, "Una opinión honrosa," *Revista del Jardín Zoológico de Buenos Aires. Época II* 4, no. 17 (April 1909): 79.
111. Zimmerman, "Racial Ideas," 34. On Ferri's influence, see Lucas Poy, "'Positive and Scientific': The Influence of Enrico Ferri in Late-Nineteenth Century Argentine Socialism," *Global Intellectual History* (September 2023): 1–19.
112. "Charlas del Pebete," *PBT* 5, no. 196 (August 15, 1908): 62.
113. Córdova [?] to Clemente Onelli, 1922, Notas de Reparticiones Nacionales, Ecoparque, Buenos Aires.
114. Karen M. Morin, *Carceral Space, Prisoners and Animals* (Abingdon: Routledge, 2018), 7.
115. *Trabajos y actas del Congreso Penitenciario Nacional celebrado en Buenos Aires del 4 al 11 de mayo de 1914* (Buenos Aires: Talleres Gráficos de la Penitenciaría Nacional, 1914), 286.
116. Ricardo D. Salvatore, *Subalternos, derechos y justicia penal: Ensayos de historia social y cultural. Argentina 1829–1940* (Barcelona: Gedisa, 2013), 215.
117. Clemente Onelli to Señor Gobernador de la Penitenciaría Nacional, December 27, 1920, Copiador 1920–1921, 174, Ecoparque, Buenos Aires.
118. Clemente Onelli to Señor Director de la Penitenciaría Nacional, March 7, 1918, Copiador 1917–1918, 442, Ecoparque, Buenos Aires.
119. Clemente Onelli, "El Jardín Zoológico y sus anexos," 431–32.
120. Eduardo Ciafardo, *Los niños en la ciudad de Buenos Aires (1890/1910)* (Buenos Aires: Centro Editor de América Latina, 1992), 50.
121. Gorelik, *The Grid and the Park*, 148.
122. Onelli, "La sucursal sur," 271.
123. Onelli, "Apuntes sobre el Jardín Zoológico," 286.
124. Onelli, "La sucursal sur," 274.
125. Clemente Onelli to Señor Secretario de la Intendencia Municipal, August 7, 1911, Copiador 1910–1912, 260, Ecoparque, Buenos Aires.
126. Onelli, "La sucursal sur," 273.
127. Clemente Onelli, "El Parque Zoológico del Sud," *Revista del Jardín Zoológico de Buenos Aires. Época II* 10, no. 39–40 (December 1914): 140.
128. J. Reybel, "Clemente Onelli," *Caras y Caretas* 21, no. 1008 (January 26, 1918).
129. Onelli, "El Parque Zoológico del Sud," 140.
130. Sandra Carreras, "'Hay que salvar en la cuna el porvenir de la patria en peligro . . .': Infancia y cuestión social en Argentina (1870–1920)," in *Entre la familia, la sociedad y el Estado: Niños y jóvenes en América Latina (siglos XIX–XX)*, ed. Barbara Potthast (Madrid: Iberoamericana, 2005), 148.
131. Clemente Onelli, "Tambos modelos y cabrerías (Memorándum pedido por el Sr. Intendente Municipal Dr. Joaquin S. de Anchorena)," *Revista del Jardín Zoológico de Buenos Aires. Época II* 7, no. 28 (December 1911): 279–81.

132. Clemente Onelli to Señor Secretario, March 25, 1921, Copiador 1920–1921, 382–384, Ecoparque, Buenos Aires.
133. Onelli, "Tambos modelos," 279.
134. Clemente Onelli, "La leche de cabra," *Revista del Jardín Zoológico de Buenos Aires. Época II* 6, no. 23 (July 1911): 196.
135. Onelli, "La leche de cabra," 197.
136. Onelli, "Tambos modelos," 289.

INTERLUDE 1

1. Clemente Onelli, "Idiosincrasias individuales de los pensionistas del Jardín Zoológico," *Revista del Jardín Zoológico de Buenos Aires. Época II* 4, no. 16 (December 1908): 334.
2. Clemente Onelli, "El peligro de las moscas," *Revista del Jardín Zoológico de Buenos Aires. Época II* 6, no. 24 (December 1910): 282.
3. Onelli, "Idiosincrasias . . ." (December 1908), 332.
4. Onelli, "Idiosincrasias . . ." (December 1908), 328.
5. Onelli, "Idiosincrasias . . ." (December 1908), 332.
6. Onelli, "Idiosincrasias . . ." (December 1908), 328.
7. Onelli, "Idiosincrasias . . ." (December 1908), 335.
8. Tarō Gomi, *Everyone Poops*, trans. Amanda Mayer Stinchecum (La Jolla: Turtleback Books, 2018).
9. Christopher L. Dutton et al., "The Meta-Gut: Community Coalescence of Animal Gut and Environmental Microbiomes," *Scientific Reports* 11, no. 1 (November 30, 2021).
10. Clemente Onelli, "Cómo se curan los ignorantes: Remedios empíricos sacados de los animales," *Revista del Jardín Zoológico de Buenos Aires. Época II* 1, no. 2 (July 1905): 139.
11. Clemente Onelli to Señor Secretario, March 25, 1921, Copiador 1920–1921, 382–384, Ecoparque, Buenos Aires.
12. Silvia Urich, *Los perritos bandidos: La protección de los animales de la Ley Sarmiento a la Ley Perón*, 2nd ed. (Temperley, Argentina: Tren en Movimiento, 2015).
13. Clemente Onelli, "Cómo se pierde y cómo se hace perder el tiempo," *Revista del Jardín Zoológico de Buenos Aires. Época II* 7, no. 25 (April 1911): 54.
14. Onelli, "Cómo se pierde," 56.
15. Onelli, "Cómo se pierde," 57–58.
16. Onelli, "Cómo se pierde," 59.
17. Vandersommers, *Entangled Encounters*, 161.
18. Onelli, "Cómo se pierde," 63.
19. Onelli, "Cómo se pierde," 86.
20. Onelli, "Cómo se pierde," 73.
21. Onelli, "Cómo se pierde," 76.

CHAPTER 2

1. Castigat ridendo mores, "Vida social zoológica," *Revista del Jardín Zoológico de Buenos Aires. Época II* 6, no. 20 (January 1910): 373.
2. I have found the aristocratic representation of (most) zoo animals in dozens of texts, including *PBT*, *Caras y Caretas*, *Fray Mocho*, *Billiken*, the *Revista del Jardín Zoológico*, *La Vanguardia*, *Crítica*, and *La Nación*.
3. "En el Palacio Novedades: El mono Cónsul, patinador, sportsman," *Caras y Caretas* 13, no. 613 (July 2, 1910).

4. Clemente Onelli, "Su alteza real la chinchilla," *Caras y Caretas* 15, no. 783 (October 19, 1912).
5. In Pedernera and Pagani, *Sarmiento*, 26.
6. Adelman, *Civilised by Beasts*, 33.
7. Ramón Columba, "Apuntes del Zoológico," *Páginas de Columba* 2, no. 7 (June 1923).
8. David Rock, *Politics in Argentina, 1890–1930: The Rise and Fall of Radicalism* (London: Cambridge University Press, 1975), 21.
9. Rock, *Politics in Argentina*, 41.
10. Lucas Poy, "Teoría y práctica del 'primer diputado socialista de América': La acción de Alfredo Palacios en el Parlamento argentino y sus repercusiones en la prensa socialista, 1904–1908," *Nuevo mundo mundos nuevos*, 2018, https://doi.org/10.4000/nuevomundo.71852.
11. Munck, *Argentina: From Anarchism*, 50.
12. Munck, *Argentina: From Anarchism*, 85.
13. Carbone, *Park, Tenement, Slaughterhouse*, 169.
14. Carbone, *Park, Tenement, Slaughterhouse*, 165–66.
15. Héctor A. Bignone, "Domingos de verano," *Caras y Caretas* 20, no. 959 (February 17, 1917).
16. Coiticher, "Picnics y fiestas campestres."
17. "En el Zoo," *La Nación*, August 16, 1905.
18. John Alexander Hammerton, *The Real Argentine: Notes and Impressions of a Year in the Argentine and Uruguay* (New York: Dodd, Mead, 1915), 111. Hammerton's reaction to the Buenos Aires Zoo is very similar to William T. Hornaday's concerns about the New York Zoological Park (today, the Bronx Zoo) where he blamed "lower-class aliens" and Italians for causing disorder. Bender, *The Animal Game*, 32.
19. Onelli, "El Jardín Zoológico en 1907," 364.
20. Diego Novillo Quiroga, "Día de Domingo" in *Sabañones*, 45–46 (Buenos Aires: Tixi & Schaffner, 1922).
21. Novillo Quiroga, *Sabañones*, 9.
22. "En el Zoo," *La Nación*, June 25, 1914.
23. Tacuabé, "Desde un banco del zoo," *La Nación*, September 15, 1906.
24. A. Regulo, "Necedades," *La Vanguardia*, September 18, 1906.
25. Augusto Vaccari, "Lo que va de la semana: Zoológicas," *Caras y Caretas* 24, no. 1202 (October 15, 1921).
26. Adolfo Lanús, "Impresiones del Zoológico: El desayuno," *Caras y Caretas* 22, no. 1103 (November 22, 1919).
27. "En el Jardín Zoológico: Protestas infantiles," *La Prensa*, May 20, 1917.
28. "Una visita al Jardín Zoológico: El sentimiento del arte en las fieras de Palermo," *Caras y Caretas* 3, no. 70 (February 3, 1900).
29. Obregón, "Congreso de Animales: Sesión extraordinaria en honor del Doctor Albarracín," *El Gladiador* 3, no. 140 (August 5, 1904).
30. Goyo Cuello, "Equitación infantil en el Zoo," *Caras y Caretas* 22, no. 1075 (May 10, 1910).
31. Rafael Sanhueza Lizardi, *Un invierno en Buenos Aires* (Paris: Paul Ollendorff, 1914), 148; Blasco Ibáñez, *Argentina y sus grandezas*, 512.
32. Gabriel Carrasco, *Luciérnagas* (Buenos Aires: Compañía Sud-Americana de Billetes de Banco, 1905).
33. Pío Cid, "Los domingos del Zoo," *PBT* 5, no. 182 (May 9, 1908): 64.
34. Julio Castellanos, "En busca de casa," *Caras y Caretas* 7, no. 358 (July 8, 1905).
35. Rock, *Politics in Argentina*, 106.

36. Manuel Redondo, "Noticias," *Caras y Caretas* 12, no. 801 (February 7, 1914). The idea that zoo animals ate better than working-class humans was not uniquely Argentine. The Irish magazine *Paddy Kelly's Budget* made the same comparison in 1835. In Adelman, *Civilised by Beasts*, 35.
37. Esteban Dagnino, "En el Jardín Zoológico: Reflexiones de un visitante," *La Vanguardia*, December 28, 1906.
38. "El Zoológico no progresa: Carta de un león," *La Razón*, August 29, 1922.
39. "Desde los dominios de Onelli: Vuelve a escribirnos un tigre de Bengala," *La Razón*, September 4, 1922.
40. Esteban Dagnino, "Parques y paseos: El Zoo," *La Vanguardia*, December 12, 1914.
41. Clemente Onelli, "El Jardín Zoológico en 1910," *Revista del Jardín Zoológico de Buenos Aires. Época II* 6, no. 24 (December 1910): 303.
42. Dagnino, "Parques y paseos."
43. "Majaderías," *La Vanguardia*, September 8, 1908.
44. "¡Sayán ha muerto! ¡Viva Sayán!" *La Vanguardia*, March 30, 1909.
45. "Mucanga burguesa," *La Vanguardia*, February 13, 1904.
46. Barrancos, *La escena iluminada*, 40.
47. "Bibliografía," *La Vanguardia*, September 6, 1906.
48. "Instrucción popular: Sociedad Luz, la excursión al Jardín Zoológico," *La Vanguardia*, October 21, 1913.
49. "Sociedad Luz: Universidad Popular," *La Vanguardia*, September 5, 1906.
50. Yaverovski, "Tres repúblicas barriales," 39.
51. Hugo Miatello, "Una institución benéfica: La Universidad Popular de La Boca," *Caras y Caretas* 22, no. 1087 (August 2, 1919).
52. Clemente Onelli, "La filosofía de la avicultura práctica: Conferencias hebdomadarias en la Universidad Popular de La Boca," *Revista del Jardín Zoológico de Buenos Aires. Época II* 13, no. 49–50 (July 1917): 85–98.
53. Clemente Onelli, "A la par con el yankee," *El Monitor de Educación Común* 36, no. 540 (December 31, 1917): 133–35.
54. Rock, *Politics in Argentina*, 21–22.
55. Clemente Onelli, "El empleado en la colmena humana: Conferencia leída la noche del 23 de septiembre de 1916 en el Club Sindical de Empleados," *Revista del Jardín Zoológico de Buenos Aires. Época II* 12, no. 47 (October 1916): 334.
56. *Anuario Social 1913* (Buenos Aires: Pablo Gadola, 1913), 185.
57. Clemente Onelli, "Desencantos e ilusiones de un naturalista (Conferencia leída en la Sociedad Luz)," *Revista del Jardín Zoológico de Buenos Aires. Época II* 12, no. 48 (December 1916): 488.
58. Onelli's predecessor, Eduardo L. Holmberg, was removed from his post after ideological and personal conflicts with the mayor of Buenos Aires. His successor, Adolfo Holmberg, lost his position after the 1943 coup d'état. Esparrach and Sassaroli, *Historia del Zoológico*, 52, 84.
59. Leandro Losada, *La alta sociedad en la Buenos Aires de la belle epoque: Sociabilidad, estilo de vida e identidades* (Buenos Aires: Siglo XXI, 2008), 308.
60. Castigat ridendo mores, "Vida social zoológica" *Revista del Jardín Zoológico de Buenos Aires. Época II* 6, no. 22 (July 1910): 142.
61. Castigat ridendo mores, "Vida social zoológica" (January 1910), 375.
62. Clemente Onelli, "No confundir el resabio con la tradición," *Revista del Jardín Zoológico de Buenos Aires. Época II* 10, no. 39–40 (December 1914): 175–90.

63. Clemente Onelli, "Buen trato hacia los animales, enfermedades que contagian, etc.: Conferencia tenida por el Director del Jardín Zoológico a los miembros de la Universidad Popular 'Luz,'" *Revista del Jardín Zoológico de Buenos Aires. Época II* 2, no. 7 (October 1906): 232.
64. Onelli, "Buen trato," 235–56.
65. Onelli, "Buen trato," 233.
66. Onelli, "Buen trato," 238.
67. Clemente Onelli, "Idiosincrasias individuales de los pensionistas del Jardín Zoológico," *Revista del Jardín Zoológico de Buenos Aires. Época II* 9, no. 34–36 (July–December 1913): 75.
68. Sophie Everest, "'Under the Skin': The Biography of a Manchester Mandrill," in *The Afterlives of Animals: A Museum Menagerie*, ed. J. M. M. Alberti (Charlottesville: University of Virginia Press, 2011), 77–78.
69. Clemente Onelli, "Idiosincrasias individuales de los pensionistas del Jardín," *Revista del Jardín Zoológico de Buenos Aires. Época II* 2, no. 6 (July 1906): 84.
70. Clemente Onelli, "Idiosincrasias individuales de los pensionistas del Jardín Zoológico," *Revista del Jardín Zoológico de Buenos Aires. Época II* 5, no. 19 (October 1909): 182.
71. Farmer's dog, "perro del hortelano," is a Spanish expression that means someone that does not want something but will not let others have it either. Clemente Onelli, "Idiosincrasias individuales de los pensionistas del Jardín Zoológico," *Revista del Jardín Zoológico de Buenos Aires. Época II* 11, no. 41 (May 1915): 13.
72. "Hazañas trágicas y ridículas de los pensionistas del Zoo," *Caras y Caretas* 9, no. 413 (September 1, 1906).
73. Clemente Onelli, "Idiosincrasias individuales de los pensionistas del Jardín Zoológico," *Revista del Jardín Zoológico de Buenos Aires. Época II* 13, no. 49–50 (July 1917): 9; Clemente Onelli, "Idiosincrasias individuales de los pensionistas del Zoológico," *Revista del Jardín Zoológico de Buenos Aires. Época II* 18, no. 66–68 (October 1922): 134.
74. Giorgi, *Formas comunes*, 15.
75. Lucas Poy, *El Partido Socialista argentino, 1896–1912: Una historia social y política* (Santiago: Ariadna, 2020), 71.
76. Poy, *El Partido Socialista*, 72.
77. Onelli, "Idiosincrasias" (July–December 1913), 74.
78. Onelli, "Idiosincrasias" (October 1907), 201.
79. "We're plowing," "estamos arando," is from a fable, and said by a fly sitting on an oxen's horns. Onelli, "Idiosincrasias" (October 1907), 203–4.
80. Onelli, "Idiosincrasias" (October 1907), 204.
81. Clemente Onelli, "La plaga de los gorriones," *Revista del Jardín Zoológico de Buenos Aires. Época II* 11, no. 42–43 (October 1915): 268.
82. Clemente Onelli, "Gorriones sovietistas," *Fray Mocho* 17, no. 828 (March 6, 1928): 38.
83. Clemente Onelli, "Idiosincrasias individuales de los pensionistas del Jardín Zoológico," *Revista del Jardín Zoológico de Buenos Aires. Época II* 6, no. 18 (July 1909): 98–99.
84. Onelli, "Idiosincrasias" (July 1909), 99.
85. Martín Fierro, "Música de actualidad," *Ideas y figuras* 1, no. 1 (May 13, 1909).
86. Munck, *Argentina: From Anarchism*, 51.
87. Cao, "La fiesta del trabajo en el Zoo," *Caras y Caretas* 9, no. 396 (May 5, 1906).
88. Alexandra Hui, Lissa Roberts, and Seth Rockman, "Introduction: Launching a Labor History of Science," *Isis* 114, no. 4 (December 2023): 822.

89. Gabriela Soto Laveaga, "Worker Once Known: Thinking with Disposable, Discarded, Mislabeled, and Precariously Employed Laborers in History of Science," *Isis* 114, no. 4 (December 2023): 837.
90. María Esther Llambias, first listed as a *peón* (laborer) and then as the second *avicultor(a)* (aviculturist). I will discuss Llambias and several other women who worked off the payroll in Chapter 4.
91. List of zoo employees, 1917, Copiador 1916–1917, Ecoparque, Buenos Aires.
92. Clemente Onelli to Señor Secretario de la Intendencia Municipal, Copiador 1916–1917, 1–2, Ecoparque, Buenos Aires.
93. Clemente Onelli to Señor Contador General de la Intendencia Municipal, March 14, 1918, Copiador 1917–1918, 459–465, Ecoparque, Buenos Aires.
94. Clemente Onelli to Señor Secretario, October 26, 1920, Copiador 1920–1921, 29, Ecoparque, Buenos Aires.
95. Clemente Onelli to Señor Secretario, May 11, 1917, Copiador 1916–1917, 272–275, Ecoparque, Buenos Aires.
96. "En el ferrocarril del Zoo," *Sherlock Holmes* 1, no. 26 (December 26, 1911): 50.
97. Departamento Nacional del Trabajo, *Boletín No. 42 del Departamento Nacional del Trabajo: Anuario estadístico 1917* (Buenos Aires: Gotelli y Cía., 1919), 11.
98. Clemente Onelli to Señor Secretario de la Intendencia Municipal, October 28, 1916, Copiador 1916–1917, 141–42, Ecoparque, Buenos Aires.
99. Clemente Onelli to Señor Secretario de la Intendencia Municipal, October 28, 1916, Copiador 1916–1917, 141–42, Ecoparque, Buenos Aires.
100. Clemente Onelli to Señor Secretario de la Intendencia Municipal, September 10, 1910. Copiador 1910–1912, 40, Ecoparque, Buenos Aires.
101. Clemente Onelli to Señor Intendente Municipal de la Capital, April 1, 1919, Copiador 1918, 1918–1919, 523, Ecoparque, Buenos Aires.
102. "Los peones del Jardín Zoológico," *La Nación*, October 22, 1918; "De las provincias," *La Vanguardia*, October 23, 1918.
103. Clemente Onelli to Señor Secretario de la Intendencia Municipal, May 23, 1917, Copiador 1916–1917, 229, Ecoparque, Buenos Aires.

INTERLUDE 2

1. *Guía popular . . . 1904*, 79.
2. Alejandro Pavlovsky to Señor Presidente de la Comisión Consultiva Dr. J. Penna, April 27, 1903, Copiador 1903, 1904, 1905, 59, Ecoparque, Buenos Aires; Clemente Onelli to Señor Secretario de la Intendencia Muncipal, June 25, 1917, Copiador 1916–1917, 311, Ecoparque, Buenos Aires.
3. Clemente Onelli to Señores Broaststael-Ubrilla, March 18, 1907, Copiador 1907–1908, 43, Ecoparque, Buenos Aires; Clemente Onelli to Señor Secretario de la Intendencia Muncipal, January 5, 1916, Copiador 1916–1917, 3, Ecoparque, Buenos Aires; Clemente Onelli to Señor Administrador General de Limpieza y Transportes, September 23, 1913, Copiador 1913–1914, 201, Ecoparque, Buenos Aires.
4. Clemente Onelli to Señor Administrador General de Limpieza y Transportes, December 18, 1913, Copiador 1913–1914, 260–261, Ecoparque, Buenos Aires.
5. Kendra Coulter, "Toward Humane Jobs and Work-Lives for Animals," in *Animal Labour: A New Frontier of Interspecies Justice?*, ed. Charlotte E. Blattner, Kendra Coulter, and Will Kymlicka (Oxford: Oxford University Press, 2020), 33.

6. Katherine Dashper, "More-Than-Human Emotions: Multispecies Emotional Labour in the Tourism Industry," *Gender, Work & Organization* 27, no. 1 (January 2020): 29.
7. Carbone, *Park, Tenement, Slaughterhouse*, 207.
8. Violette Pouillard, "The Silence and the Fury: Addressing Animal Resistance and Agency through the History of Human-Animal Relationships," in *Animals Matter: Resistance and Transformation in Animal Commodification*, ed. Julien Dugnoille and Elizabeth Vander Meer (Leiden: Brill, 2023), 45.
9. Jason Hribal, "Animals, Agency, and Class: Writing the History of Animals from Below," *Human Ecology Review* 14, no. 1 (Summer 2007): 103.
10. Martín Fierro, hijo, "Jardín Zoológico (Palermo)," *Iris* 1, no. 17 (November 23, 1899).
11. "Carnaval: El primer día de fiestas," *La Prensa*, February 19, 1917.
12. Juan Carlos Huergo, "Un domingo en el Zoo," *La Nación*, August 14, 1921.
13. Clemente Onelli, "Idiosincrasias individuales de los pensionistas del Jardín," *Revista del Jardín Zoológico de Buenos Aires. Época II* 4, no. 13 (April 1908): 10.
14. "En el Jardín Zoológico: Búfalo enfurecido," *La Prensa*, October 22, 1911.
15. Clemente Onelli, "Idiosincrasias individuales de los pensionistas del Jardín Zoológico," *Revista del Jardín Zoológico de Buenos Aires. Época II* 11, no. 44 (December 1915): 332.
16. "La comida de las fieras," *Fray Mocho* 2, no. 7 (October 3, 1913).
17. "Hazañas trágicas," *Caras y Caretas*.
18. Onelli, "Idiosincrasias" (December 1908), 344.
19. Onelli, "Idiosincrasias" (October 1909), 196.
20. Clemente Onelli, "Idiosincrasias individuales de los pensionistas del Jardín Zoológico," *Revista del Jardín Zoológico de Buenos Aires. Época II* 6, no. 21 (April 1910): 10–11.
21. Onelli, "Idiosincrasias" (April 1910), 12.
22. Hribal, *Fear of the Animal Planet*, 114.
23. Adelman, *Civilised by Beasts*, 4.
24. One study of the relationship between zoo animals and a zoo employee in Argentina is Jessica Sarah Rinland's 2024 film *Monólogo colectivo*, which intertwines the relationship between the keeper Macarena and the creatures she cares for at the La Plata zoo with a broader examination of the history of zoos and contemporary animal rescue sites. I had the privilege of meeting and working with the filmmaker at the Ecoparque Archive in 2022.
25. *Guía popular . . . 1904*, 42.
26. Alasdair Cochrane, "Good Work for Animals," in *Animal Labour: A New Frontier of Interspecies Justice?*, ed. Charlotte E. Blattner, Kendra Coulter, and Will Kymlicka (Oxford: Oxford University Press, 2020), 49.
27. Jocelyne Porcher, "The Relationship Between Workers and Animals in the Pork Industry: A Shared Suffering," *Journal of Agricultural and Environmental Ethics* 24, no. 1 (February 2011): 6.
28. Clemente Onelli, "Los derechos del hombre y del animal," *Revista del Jardín Zoológico de Buenos Aires. Época II* 7, no. 27 (October 1911): 230.
29. Onelli, "Idiosincrasias" (May 1915), 27.

CHAPTER 3

1. Manuel Redondo, "Sarrasqueta busca un perro," *Caras y Caretas* 12, no. 801 (February 7, 1914).
2. Nicola Miller, *In the Shadow of the State: Intellectuals and the Quest for National Identity in Twentieth-Century Spanish America* (New York: Verso, 1999), 164.

3. Scarzanella, *Ni gringos ni indios*, 30.
4. Carlos Altamirano and Beatriz Sarlo, "La Argentina del Centenario: Campo intelectual, vida literaria y temas ideológicos," in *Ensayos argentinos: de Sarmiento a la vanguardia*, ed. Carlos Altamirano and Beatriz Sarlo (Buenos Aires: Siglo Veintiuno Editores, 2016), 156.
5. See Jean Delaney, "Imagining 'El Ser Argentino': Cultural Nationalism and Romantic Concepts of Nationhood in Early Twentieth-Century Argentina," *Journal of Latin American Studies* 34, no. 3 (August 2002): 625–58; Miller, *In the Shadow of the State*.
6. Manuel Gálvez, *El solar de la raza*, 5th ed. (Madrid: Saturnino Calleja, 1920), 17–18.
7. Flack, *The Wild Within*, 67.
8. In Pedernera and Pagani, *Sarmiento*, 47.
9. The records of donations encompass 1875–1887, then 1898–1904. *Guía popular . . . 1904*, 22–28.
10. Ralph, "Tatú-Carreta, del Chaco," *Caras y Caretas* 16, no. 778 (August 30, 1913).
11. Gastón Gordillo and Silvia Hirsch, "Indigenous Struggles and Contested Identities in Argentina: Histories of Invisibilization and Reemergence," *Journal of Latin American Anthropology* 8, no. 3 (2003): 11.
12. Miller, *In the Shadow*, 169.
13. Erika Denise Edwards, *Hiding in Plain Sight: Black Women, the Law, and the Making of a White Argentine Republic* (Tuscaloosa: University of Alabama Press, 2021), 3.
14. Clemente Onelli to Monsieur S. Gardette, October 15, 1911, Copiador 1910–1912, 203, Ecoparque, Buenos Aires.
15. "La girafa 'Mimí,'" *Caras y Caretas*.
16. Armus, *The Ailing City*, 229.
17. "La jirafa del Zoo," *La Semana Universal* 1 (May 2, 1912).
18. "La girafa 'Mimí,'" *Caras y Caretas*.
19. *Guía oficial . . . 1920–1921*, 21.
20. "Menudencias," *Caras y Caretas* 15, no. 709 (May 4, 1912).
21. Hanson, *Animal Attractions*, 45.
22. Pavlovsky, "Sección administrativa," 316.
23. Clemente Onelli, "Nuevas adquisiciones," *Revista del Jardín Zoológico de Buenos Aires. Época II* 1, no. 3 (October 1905): 255.
24. Clemente Onelli, "Jardín Zoológico del Sur," *Revista del Jardín Zoológico de Buenos Aires. Época II* 17, no. 65 (April 1921): 115.
25. My calculation, based on the species inventories in the 1904 and 1920–1921 zoo guidebooks.
26. Ralph, "Tatú-Carreta, del Chaco."
27. Charles Warren Currier, *Lands of the Southern Cross: A Visit to South America* (Washington, DC: Spanish-American Publication Society, 1911), 113–14.
28. Regina Horta Duarte, "'El zoológico del porvenir': Narrativas y memorias de nación sobre el Zoológico de Chapultepec, Ciudad de México, siglo XX," *Historia Crítica* 72 (2019): 101.
29. Juan de Cominges, "Palermo y el Zoológico hasta 1882," *Revista del Jardín Zoológico de Buenos Aires. Época II* 12, no. 45 (May 1916): 78.
30. Eduardo L. Holmberg, "El Jardín Zoológico en 1890: La primera guía del Jardín Zoológico de Buenos Aires," *Revista del Jardín Zoológico de Buenos Aires* 1, no. 8 (August 15, 1893): 235.
31. Holmberg, "El Jardín Zoológico en 1890," 235–236.

32. Clemente Onelli, "Zoología útil: Fauna que desaparece, su protección, las alpacas," *Revista del Jardín Zoológico de Buenos Aires. Época II* 1, no. 3 (October 1905): 241.
33. Leopoldo Lugones, *Didáctica* (Buenos Aires: Otero, 1910), 270.
34. Carlos Octavio Bunge, *Nuestra patria: Libro de lectura para la educación nacional*, 24th ed. (Buenos Aires: Estrada, 1910).
35. Ricardo Rojas, *La restauración nacionalista: Informe sobre educación* (Buenos Aires: Ministerio de Justicia e Instrucción Pública, 1909), 65.
36. Warwick Anderson, "Climates of Opinion: Acclimatization in Nineteenth-Century France and England," *Victorian Studies* 35, no. 2 (Winter 1992): 135–36.
37. Harriet Ritvo, "The World as Zoo: Acclimatization in the Nineteenth Century," in *The Ark and Beyond: The Evolution of Zoo and Aquarium Conservation*, ed. Ben A. Minteer, Jane Maienschein, and James P. Collins (Chicago: University of Chicago Press, 2018), 32.
38. Ritvo, "The World as Zoo," 36.
39. "Liste supplémentaire des membres de la société," *Bulletin de la Société Nationale d'Acclimatation de France* 58 (January 1, 1911).
40. Helen Cowie, "From the Andes to the Outback: Acclimatising Alpacas in the British Empire," *Journal of Imperial and Commonwealth History* 45, no. 4 (July 4, 2017): 566.
41. Pascal Grosse, "Turning Native?: Anthropology, German Colonialism, and the Paradoxes of the 'Acclimatization Question,' 1885–1914," in *Worldly Provincialism: German Anthropology in the Age of Empire*, ed. H. Glenn Penny and Matti Bunzi (Ann Arbor: University of Michigan Press, 2003), 181.
42. Peter Maxwell Minard, *All Things Harmless, Useful, and Ornamental: Environmental Transformation through Species Acclimatization, from Colonial Australia to the World* (Chapel Hill: University of North Carolina Press, 2019).
43. Clemente Onelli, "Ganadería exótica," *Anales de la Sociedad Rural Argentina* 40 (May–June 1906): 66.
44. Augusto Huber, "El karakul," *Revista del Jardín Zoológico de Buenos Aires. Época II* 8, no. 29 (April 1912): 53.
45. Huber, "El karakul," 57.
46. Huber, "El karakul," 59.
47. Clemente Onelli, "El karakul criollo," *Revista del Jardín Zoológico de Buenos Aires. Época II* 8, no. 30 (July 1912): 123–24.
48. Onelli, "El karakul criollo," 128.
49. Fernando Devoto, *Historia de la inmigración en la Argentina*, 2nd ed. (Buenos Aires: Sudamericana, 2004), 298.
50. Clemente Onelli, "Idiosincrasias individuales de los pensionistas del Jardín," *Revista del Jardín Zoológico de Buenos Aires. Época II* 3, no. 9 (April 1907): 13.
51. The text was published in installments in the *Revista* and was available for purchase at the zoo for $2 per pamphlet. It was meant to be a nationalist teaching tool, focusing on the biology of the region and able to be used at all education levels. Christofredo Jakob, "Tratado de biología general y especial para el uso de la enseñanza elemental, secundaria y superior en la República Argentina," *Revista del Jardín Zoológico de Buenos Aires. Época II* 11, no. 42–43 (October 1915): 275–277.
52. Christofredo Jakob, "Tratado de biología general y especial: Biología vegeto-animal elemental," *Revista del Jardín Zoológico de Buenos Aires. Época II* 12, no. 48 (December 1916): 553.
53. Jakob, "Tratado de biología . . .: Biología vegeto-animal," 684.

54. Robert Lehmann-Nitsche, "Aclimatación de la perdiz grande y de la martineta en Alemania," *Revista del Jardín Zoológico de Buenos Aires. Época II* 3, no. 10 (July 1907): 132.
55. "Una riqueza rural que se desdeña," *Anales de la Sociedad Rural Argentina* 38–39 (1904): 41.
56. Clemente Onelli, "El avestruz criollo en moda," *Anales de la Sociedad Rural Argentina*, no. 42–43 (1908): 71.
57. Onelli, "El avestruz criollo," 72.
58. Clemente Onelli, "Ganadería de pelo fino," *Anales de la Sociedad Rural Argentina* 42–43 (1908): 62. Relatedly, Onelli served on a committee convened by President Sáenz Peña to discuss the fate of the guanaco, a creature that livestock owners in the south sought to have declared a "national plague." Onelli argued against this proposal, insisting instead on the development of a guanaco fur industry to control the population. John Soluri, *Creatures of Fashion: Animals, Global Markets, and the Transformation of Patagonia* (Chapel Hill: University of North Carolina Press, 2024), 138.
59. Onelli, "El avestruz criollo," 73.
60. Lehmann-Nitsche, "Aclimatación," 132.
61. Lucas Poy, "Ciudadanía, derechos políticos y conciencia de clase: La cuestión de la naturalización de los extranjeros en los orígenes del socialismo argentino," *Diálogos: Revista Electrónica de Historia* 16, no. 2 (December 2015): 195.
62. Poy, "Ciudadanía, derechos políticos," 203.
63. María Inés Tato, "Del crisol de razas a la Argentina desintegrada: Un itinerario de la idea de nación, 1911–1932," *Historia y Política* 17 (June 2007): 165.
64. Clemente Onelli, "Sociología-Zoológica," *Caras y Caretas* 15, no. 724 (August 17, 1912).
65. "Correo sin estampilla," *Caras y Caretas* 23, no. 1120 (March 20, 1920).
66. Donna J. Guy, *Sex and Danger in Buenos Aires: Prostitution, Family, and Nation in Argentina* (Lincoln: University of Nebraska Press, 1991), 2.
67. Eduardo L. Holmberg, "Zoológico de Buenos Ayres: Mamíferos. Cueros armados existentes en el museo del Jardín el día 31 de diciembre de 1892," *Revista del Jardín Zoológico de Buenos Aires* 1, no. 4 (April 15, 1893): 116.
68. Onelli, "Sociología-Zoológica."
69. "El acontecimiento zoo-cial de la semana," *Caras y Caretas* 9, no. 404 (June 30, 1906).
70. Onelli, "Una mirada retrospectiva . . . 1906," 310.
71. *Guía popular . . . 1904*, 87.
72. Arthur Ruhl, *The Other Americans: The Cities, the Countries, and Especially the People of South America* (New York: Scribner's Sons, 1910), 214.
73. "Menudencias," *Caras y Caretas* 9, no. 387 (March 3, 1906).
74. "Charlas del Pebete," *PBT*, 62.
75. Devoto, *Historia de la inmigración*, 277.
76. Carlos Correa Luna, "El espíritu de rebelión," *Caras y Caretas* 7, no. 300 (July 2, 1904).
77. Obregón, "Congreso de Animales."
78. Holmberg, "El Jardín Zoológico en 1890," 253.
79. *Guía oficial . . . 1920–1921*, 81–82.
80. Onelli, "Idiosincrasias" (April 1908), 11.
81. Onelli, "Idiosincrasias" (April 1908), 12.
82. Clemente Onelli, "La étnica social hebrea: Conferencia pronunciada el 30 de junio de 1917 en el Centro Juventud Israelita," *Revista del Jardín Zoológico de Buenos Aires. Época II* 13, no. 49–50 (July 1917): 70.
83. Onelli, "La étnica social hebrea," 80.

84. Onelli, "La étnica social hebrea," 81.
85. Onelli, "La étnica social hebrea," 82.
86. Onelli, "La étnica social hebrea," 83.
87. Onelli, "La étnica social hebrea," 84. Onelli's relationship with Buenos Aires's large Jewish community is worth a study of its own. When he died, *Mundo Israelita* published an obituary stating that he was very popular "among the Jews" and had given numerous speeches to Jewish cultural centers, published articles in Jewish magazines, and had many Jewish good friends. The piece also acknowledged the tension between these close relationships and Onelli's statements on Jewish assimilation, saying, "and although the popular Zoological Garden director has not always treated our problems and aspirations with the justice we think they deserve, we know him to be of good faith and without preconceived notions." "Clemente Onelli," *Mundo Israelita*, October 25, 1924.
88. Clemente Onelli, "El sionismo integral," *Vida Nuestra* 1, no. 12 (June 1918): 286.
89. Onelli, "El sionismo integral," 287.
90. Rusiñol, *Un viaje al Plata*, 171.
91. Libedinsky, "La excursión escolar," 175.
92. Antonio Prado, "El alsaciano y el orador del Zoo," *PBT* 9, no. 371 (January 6, 1912): 97.
93. For examples of zoos and imperialism, see Baratay and Hardouin-Fugier, *Zoo*, 125–29. For a more theoretical take on zoos and imperialism, see Malamud, *Reading Zoos*, 59–65.
94. Juan Pablo Scarfi, "La emergencia de un imaginario latinoamericanista y antiestadounidense del orden hemisférico: De la Unión Panamericana a la Unión Latinoamericana (1880–1913)," *Revista Complutense de Historia de América* 39 (January 9, 2014): 89.
95. Mark Petersen, *The Southern Cone and the Origins of Pan America, 1888–1933* (Notre Dame, IN: University of Notre Dame Press, 2022), 2. Petersen's book is a detailed analysis of Argentina's many approaches to pan-Americanism. Other useful texts include Richard Cándida Smith, *Improvised Continent: Pan-Americanism and Cultural Exchange* (Philadelphia: University of Pennsylvania Press, 2017), and Juan Pablo Scarfi and David Sheinin, eds., *The New Pan-Americanism and the Structuring of Inter-American Relations* (New York: Routledge, Taylor & Francis Group, 2022).
96. Interestingly, Gálvez's most famous character, Nacha Regules, goes to the zoo. See Chapter 1 for more.
97. Michela Coletta, *Decadent Modernity: Civilization and "Latinidad" in Spanish America, 1880–1920* (Liverpool: Liverpool University Press, 2018), 31.
98. Scarfi, "La emergencia," 90.
99. "Menudencias," *Caras y Caretas* 9, no. 401 (June 8, 1906).
100. Petersen, *The Southern Cone*, 79–82.
101. "El zoológico social," *PBT* 5, no. 201 (September 19, 1908): 125.
102. Clemente Onelli, "Idiosincrasias de los pensionistas del Jardín," *Revista del Jardín Zoológico de Buenos Aires. Época II* 2, no. 7 (October 1906): 182.
103. Onelli, "Idiosincrasias" (October 1906), 184.
104. Roosevelt and his son Kermit visited Buenos Aires in November 1913. *Caras y Caretas* documented their stay, including their visit to the zoo. Onelli gave them a personal tour that involved riding the miniature train and petting several animals. Given Onelli's aversion to US imperialism, it would be fascinating to know what the two men spoke about. "La visita de Mr. Roosevelt," *Caras y Caretas* 16, no. 789 (November 15, 1913).
105. Clemente Onelli, "Soplando en la llama de 'Ideas,'" *Revista del Jardín Zoológico de Buenos Aires. Época II* 14, no. 54–55 (July 1918): 183.

106. Clemente Onelli, "Ideas disonantes sobre educandos y educacionistas," *Revista del Jardín Zoológico de Buenos Aires. Época II* 9, no. 34–36 (December 1913): 184.
107. Manuscript of speech by Clemente Onelli at the Club Francés. Fondo Clemente Onelli, Sala 7, Tomo 1, 2552, Archivo General de la Nación, Buenos Aires.
108. Alexandra Pita González, *La Unión Latino Americana y el boletín "Renovación": Redes intelectuales y revistas culturales en la década de 1920* (México, DF: El Colegio de México, Centro de Estudios Históricos, 2009), 18–19.
109. Patricia V. Corsani, "La tarea de ornamentar plazas y paseos: El origen de las primeras esculturas del Jardín Zoológico," in *El pórtico bizantino del Jardín Zoológico de Buenos Aires: Una reflexión sobre nosotros mismos*, ed. Liliana Barela (Buenos Aires: Dirección General Patrimonio e Instituto Histórico de la Ciudad de Buenos Aires, 2013), 100–101.
110. *Guía oficial . . . 1920–1921*, 94–95.
111. Darío himself once visited the Buenos Aires Zoo and later sent a reporter from his *Mundial Magazine* to interview Onelli. Edmundo Montagne, "El viaje de 'Mundial,'" *Mundial Magazine* 2, no. 19 (November 1912): 603.
112. Onelli, "El Parque Zoológico del Sud," 146–47.
113. Onelli, "El Parque Zoológico del Sud," 143.
114. Onelli, "El Parque Zoológico del Sud," 149.
115. These advertisements can be found in the 1920–1921 zoo guidebook.
116. Clemente Onelli, "La fauna de la América Latina," *Caras y Caretas* 16, no. 744 (January 4, 1913).
117. Ángel Bueno, "Tipos populares," *Caras y Caretas* 16, no. 744 (January 4, 1913).
118. Roque Sáenz Peña, "Resurgere," *Caras y Caretas* 16, no. 744 (January 4, 1913).
119. Onelli, "La fauna de la América Latina."
120. Ariela Borgogno, "*Revista Idea Latina*: Una mirada a la traducción cultural en los espacios de la diversidad," *Transfer*, no. 2 (November 2010): 48.
121. Borgogno, "*Revista Idea Latina*," 56.
122. Clemente Onelli, "Las Geórgicas," *Idea Latina* (1924): 5–6.
123. In J. B. Albertoni, "La Argentina país de conquista," *Estudios* 168 (June 1925): 428–429.
124. In Albertoni, "La Argentina," 430.
125. In Albertoni, "La Argentina," 430.
126. Enrique Osés, "La penetración yanqui de la que nadie habla siquiera," *Crisol*, June 25, 1939.
127. Roberto Orofino Púrpora, *Por la unidad de Hispano América* (Córdoba: Martín Fierro, 1951).
128. List of top employees, Copiador 1916–1917, Ecoparque, Buenos Aires
129. R. Schulz, "Clemente Onelli," *Boletín de la "Sarmiento"* 15, no. 137–138 (December 1924): 5.

INTERLUDE 3

1. Devoto, *Historia de la inmigración*, 247.
2. Castigat ridendo mores, "Vida social zoológica," *Revista del Jardín Zoológico de Buenos Aires. Época II* 6, no. 24 (December 1910): 290–91.
3. Onelli, "Sección médica," 94.
4. Onelli, "Idiosincrasias" (April 1908), 18.
5. "Petronio: El nuevo pensionista," *La Nación*, October 6, 1907. Apes in captivity around the world provoked intense interest and disgust, as well as fodder for debates about evolution, species boundaries, and humans' relationships to animals. Richard Burkhardt

Jr., "An Orangutan in Paris: Pondering Proximity at the Muséum d'Histoire Naturelle in 1836," *History and Philosophy of the Life Sciences* 40, no. 1 (March 2018): 2–3.

6. "Un reportaje al orangután 'Petronio,'" *Caras y Caretas* 10, no. 471 (October 12, 1907).
7. Silvia Sebastiani, "A 'Monster with Human Visage': The Orangutan, Savagery, and the Borders of Humanity in the Global Enlightenment," *History of the Human Sciences* 32, no. 4 (October 2019): 90.
8. Clemente Onelli, "La enfermedad del orangután y la coprología," *Revista del Jardín Zoológico de Buenos Aires. Época II* 3, no. 12 (December 1907): 335.
9. Onelli, "La enfermedad del orangután," 336.
10. Petronio's experience was not unique. Dinah the gorilla's first years at the New York Zoological Park were very similar to Petronio's, including gastrointestinal issues and a belief that the ape was sad. Tracy McDonald, "Sculpting Dinah with the Blunt Tools of the Historian," in *Zoo Studies: A New Humanities*, ed. Tracy McDonald and Daniel Vandersommers (Montreal: McGill-Queen's University Press, 2019), 102. Many other apes suffered similar fates.
11. Clemente Onelli, "Idiosincrasias individuales de los pensionistas del Jardín Zoológico," *Revista del Jardín Zoológico de Buenos Aires. Época II* 3, no. 12 (December 1907): 315–16.
12. Onelli, "Idiosincrasias" (December 1907), 317.
13. "Petronio," *La Nación*, January 2, 1908.
14. "El Jardín Zoológico de duelo: Muerte de Petronio," *La Prensa*, January 7, 1908.
15. "La muerte de Petronio," *La Nación*, January 7, 1908.
16. Onelli, "Idiosincrasias" (April 1908), 16–18.
17. Clemente Onelli, "Necropsia de un orangután," *Revista del Jardín Zoológico de Buenos Aires. Época II* 4, no. 13 (April 1908): 99.
18. Vandersommers, *Entangled Encounters*, 71.
19. Onelli, "Idiosincrasias" (May 1915), 26–27.
20. Onelli, "Los derechos del hombre," 231–33.
21. Onelli, "Los derechos del hombre," 230.
22. Onelli, "Los derechos del hombre," 231–33.
23. Marcus Humo, "Cosas del año nuevo," *Caras y Caretas* 11, no. 484 (January 11, 1908).

CHAPTER 4

1. Onelli, "Una mirada retrospectiva . . . 1906," 310.
2. Onelli, "La elefantita americana," 15.
3. Onelli, "Una mirada retrospectiva . . . 1906," 310.
4. Eliseo Canton, *Protección a la madre y al hijo: Puericultura intra- y extrauterina* (Buenos Aires: Hunzinger, 1913), 16.
5. Canton, *Protección a la madre*, 25.
6. Holmberg, "Informe anual," 17.
7. Onelli, "La elefantita americana," 16.
8. Clemente Onelli, "Idiosincrasias individuales de los pensionistas del Jardín," *Revista del Jardín Zoológico de Buenos Aires. Época II* 1, no. 3 (October 1905): 225.
9. Clemente Onelli, "Idiosincrasias individuales de los pensionistas del Jardín," *Revista del Jardín Zoológico de Buenos Aires. Época II* 2, no. 5 (April 1906): 7.
10. Onelli, "Idiosincrasias" (April 1906), 4–9.
11. Ashley Elizabeth Kerr, *Sex, Skulls, and Citizens: Gender and Racial Science in Argentina (1860–1910)* (Nashville, TN: Vanderbilt University Press, 2020), 25.
12. Castigat ridendo mores, "Vida social zoológica" (July 1910), 142.

13. *Guía popular . . . 1904*, 42. Clemente Onelli, "Idiosincrasias individuales de los pensionistas del Jardín Zoológico," *Revista del Jardín Zoológico de Buenos Aires. Época II* 10, no. 37–38 (July 1914): 4.
14. Lavrin, *Women, Feminism*, 227.
15. Catriona Mortimer-Sandilands, "Unnatural Passions?: Notes Toward a Queer Ecology," *Invisible Culture* 9 (2005).
16. Clemente Onelli, "La Arcadia Porteña," *Caras y Caretas* 21, no. 1009 (February 2, 1918).
17. *Guía oficial municipal del Jardín Zoológico 1906–1907* (Buenos Aires: Vasco Fogli, 1906), 43.
18. "El pequeño dromedario del Zoológico," *PBT* 46, no. 360 (October 21, 1911).
19. "Novedades del Zoo," *Sherlock Holmes* 1, no. 16 (October 17, 1911): 5.
20. Marianna Szczygielska, "Animal Sex in Public: Warping Time and Sexuality in the Zoo," *Environmental Humanities* 14, no. 3 (November 1, 2022): 646.
21. Marisa Miranda, "Maternidad y biopolítica en la Argentina: Gregorio Aráoz Alfaro, *El Libro de las Madres* y la eugenesia (1870–1955)," *Passagens: Revista Internacional de História Política e Cultura Jurídica* 11, no. 2 (August 2019): 158.
22. El Judío Errante, "Baedeker Humorístico de la República Argentina: El Botánico y el Zoológico," *PBT* 10, no. 453 (August 2, 1913).
23. Lauren Berlant and Michael Warner, "Sex in Public," *Critical Inquiry* 24, no. 2 (Winter 1998): 549.
24. Berlant and Warner, "Sex in Public," 553.
25. Castigat ridendo mores, "Vida social zoológica" (July 1910), 142–45.
26. Clemente Onelli, "Idiosincrasias individuales de los pensionistas del Jardín Zoológico," *Revista del Jardín Zoológico de Buenos Aires. Época II* 14, no. 54–55 (July 1918): 139.
27. Clemente Onelli, "Idiosincrasias individuales de los pensionistas del Jardín Zoológico," *Revista del Jardín Zoológico de Buenos Aires. Época II* 4, no. 15 (September 1908): 216.
28. Onelli, "Idiosincrasias" (May 1915), 12.
29. M. González Herrera, "Causas de muerte en los Jardines Zoológicos (continuación)," *Revista del Jardín Zoológico de Buenos Aires. Época II* 1, no. 4 (December 1905): 343–44.
30. González Herrera, "Causas de muerte," 345–346.
31. Clemente Onelli, "Algunas correspondencias cangeadas en el trimestre," *Revista del Jardín Zoológico de Buenos Aires. Época II* 4, no. 13 (April 1908): 104.
32. Clemente Onelli, "Meditando sobre la ley de Mendel: Un caso raro no previsto por ella," *Revista del Jardín Zoológico de Buenos Aires. Época II* 4, no. 14 (July 1908): 187.
33. Ángel Gallardo, "Las excepciones al Mendelismo: Una carta del Dr. A. Gallardo," *Revista del Jardín Zoológico de Buenos Aires. Época II* 4, no. 15 (September 1908): 306.
34. Marcela Nari, *Políticas de maternidad y maternalismo político: Buenos Aires, 1890–1940* (Buenos Aires: Biblos, 2004), 36.
35. Clemente Onelli, "La superioridad del animal sobre el hombre (Conferencia pronunciada por el director del J. Zoológico en la fiesta anual de la Sociedad 'Sarmiento,' protectora de animales)," *Revista del Jardín Zoológico de Buenos Aires. Época II* 10, no. 37–38 (July 1914): 58–59.
36. Gustavo Vallejo, "La hora cero de la eugenesia en la Argentina: Disputas e ideologías en el surgimiento de un campo científico, 1916–1932," *História, Ciências, Saúde-Manguinhos* 25, suppl. 1 (August 2018): 19. On the specific role of marriage and reproduction in Argentine eugenics, see Marisa Miranda, "Matrimonio y procreación en la ortodoxia eugénica argentina," *Sociohistórica* 17–18 (first and second semester 2005): 151–78.
37. Onelli, "La superioridad del animal," 58.

38. Grant authored the racist tract *The Passing of the Great Race*; Osborn wrote a preface for Grant's book. Both would go on to be founders of the American Eugenics Society (1926). For more on eugenics at the Bronx Zoo, see Katherine McLeod, "The Call of the Hoatzin: Ecology, Evolution, and Eugenics at the Bronx Zoo," *Centaurus* 64, no. 3 (October 2022): 683–704.
39. Ana Laura Martín, "Partear y cuidar en Buenos Aires (1877–1920): Una aproximación comparativa," *Anuario del Instituto de Historia Argentina* 18, no. 1 (June 2018).
40. Clemente Onelli to Señor Secretario, October 20, 1920, Copiador 1920–1921, 11, Ecoparque, Buenos Aires.
41. Nari, *Políticas de maternidad*, 53. See also Miranda, "Maternidad y biopolítica."
42. *Guía oficial . . . 1906–1907*, 185.
43. *Guía oficial . . . 1906–1907*, 205.
44. Onelli, "Idiosincrasias" (April 1908), 15.
45. "Charlas del Pebete," *PBT*, 62.
46. "El pequeño dromedario del Zoológico," *PBT*.
47. Nari, *Políticas de maternidad*, 129.
48. Clemente Onelli, "El afecto paternal y el de la madre," *Revista del Jardín Zoológico de Buenos Aires. Época II* 3, no. 9 (April 1907): 56.
49. *Guía popular . . . 1904*, 42.
50. Clemente Onelli, "'Das Ewigweiblich' o La esfinge sin enigmas," *Revista del Jardín Zoológico de Buenos Aires. Época II* 10, no. 39–40 (December 1914): 195.
51. Onelli, "'Das Ewigweiblich,'" 198.
52. Miranda, "Maternidad y biopolítica," 159.
53. Onelli, "La superioridad del animal," 54.
54. Nari, *Políticas de maternidad*, 130.
55. Onelli, "El afecto paternal," 52.
56. Onelli, "El afecto paternal," 54.
57. *Guía popular . . . 1904*, 43–44.
58. Onelli, "El afecto paternal," 53.
59. Onelli, "Idiosincrasias" (September 1908), 224.
60. Onelli, "Idiosincrasias" (September 1908), 225–26.
61. Onelli, "Idiosincrasias" (September 1908), 224; El del verde gabán, "Visita de Duelo," *PBT* 5, no. 196 (August 15, 1908): 64.
62. Clemente Onelli to Señor Secretario de la Intendencia Municipal, August 21, 1917, Copiador 1916–1917, 377, Ecoparque, Buenos Aires.
63. Clemenceau, *Notas de viaje*, 43.
64. "Jardín Zoológico: El día de un director," *Caras y Caretas* 15, no. 709 (May 4, 1912).
65. Onelli, "Idiosincrasias" (October 1907), 209; *Guía oficial . . . 1908–1909*, 25; Clemente Onelli, "Nuestro Zoo según 'La Nature,'" *Revista del Jardín Zoológico de Buenos Aires. Época II* 8, no. 31 (October 1912): 267–71.
66. Clemente Onelli, "La princesa y el pingüino," *Revista del Jardín Zoológico de Buenos Aires. Época II* 7, no. 26 (July 1911): 151.
67. "Jardín Zoológico: El día."
68. Payroll, November 1917, Copiador 1917–1918, 107–14, Ecoparque, Buenos Aires.
69. Payroll, January 1918, Copiador 1917–1918, 219–26, Ecoparque, Buenos Aires.
70. Clemente Onelli to Señor Secretario de la Intendencia Municipal, March 15, 1919, Copiador 1918, 1918–1919, 479, Ecoparque, Buenos Aires.

71. Clemente Onelli to Señor Secretario de la Intendencia Municipal, June 20, 1919, Copiador 1918, 1918–1919, 764, Ecoparque, Buenos Aires.
72. Clemente Onelli to Señor Secretario de la Intendencia Municipal Dr. Carlos Saavedra Lamas, July 15, 1907, Copiador 1907–1908, 125–26, Ecoparque, Buenos Aires.
73. Bender, *The Animal Game*, 180.
74. Nari, *Políticas de maternidad*, 57.
75. Reybel, "Clemente Onelli."
76. Cao, "Caricaturas contemporáneas: Clemente Onelli," *Caras y Caretas* 11, no. 529 (November 21, 1908).
77. Juandrea M. Bates, "Unaccompanied Minors and Fraudulent Fathers: Civil Law in the Unmaking of Immigrant Family in Buenos Aires, 1869–1920," *Hispanic American Historical Review* 102, no. 1 (February 1, 2022): 103, https://doi.org/10.1215/00182168-9497213.
78. Gregorio Aráoz Alfaro, *El libro de las madres: Manual práctico de higiene del niño, con indicaciones sobre el embarazo, parto y tratamiento de los accidentes*, Nueva edición reformada (Buenos Aires: Cabaut y Cía, 1922). Marisa Miranda astutely analyzes Aráoz Alfaro's text and the changes in subsequent editions in her "Maternidad y biopolítica."
79. Cecilia L. Allemandi, "Las amas de leche y la regulación del mercado de la lactancia en la ciudad de Buenos Aires (1875–1911)," *Mora* 22 (2016): 7.
80. Rodríguez, *Civilizing Argentina*, 117.
81. Allemandi, "Las amas de leche," 14.
82. Club de Madres, *Memoria de la 1a Semana del Nene organizada por el Club de Madres para contribuir a la disminución de la mortalidad infantil* (Buenos Aires: Wiebeck, Turtl, 1918); Nari, *Políticas de maternidad*, 119.
83. *Guía oficial . . . 1920–1921*, 120.
84. Miranda, "Maternidad y biopolítica," 159.
85. Onelli, "La superioridad del animal," 55.
86. Ski, "Los niños y los paquidermos," *Caras y Caretas* 15, no. 721 (July 27, 1912).
87. *Guía oficial . . .* 1908–1909, 5.
88. *Guía oficial . . .* 1908–1909, 6.
89. Aleksandra Wesolowska et al., "Barley Malt-Based Composition as a Galactagogue: A Randomized, Controlled Trial in Preterm Mothers," *Ginekologia Polska* 92, no. 2 (February 26, 2021): 122.
90. *Guía oficial . . .* 1906–1907, 26.
91. Club de Madres, *Memoria de la 1a Semana*, 27.
92. Club de Madres, *Memoria de la 1a Semana*, 28.
93. Onelli, "El Jardín Zoológico en 1910," 303–4.
94. Clemente Onelli, "Ameghino en la ciencia y en la vida privada," *Revista del Jardín Zoológico de Buenos Aires. Época II* 11, no. 42–43 (October 1915): 233.
95. Onelli, "Ameghino en la ciencia," 232.
96. Clemente Onelli, "Escuela municipal de telares domésticos," *Revista del Jardín Zoológico de Buenos Aires. Época II* 18, no. 66–68 (October 1922): 163–164.
97. Nari, *Políticas de maternidad*, 85.
98. Marcela María Alejandra Nari, "El feminismo frente a la cuestión de la mujer en las primeras décadas del siglo XX," in *La cuestión social en la Argentina, 1870–1943*, ed. Juan Suriano (Buenos Aires: La Colmena, 2000), 292.
99. Onelli, "Escuela municipal," 165.
100. Onelli, "Escuela municipal," 166.
101. Onelli, "Escuela municipal," 163.

102. Andrea S. Pegoraro, "El uso de motivos indígenas de colecciones del Museo Etnográfico de la UBA en los inicios del siglo XX: Actores, actividades y objetos," *Revista del Museo de Antropología* 10, no. 1 (June 2017): 32.
103. Clemente Onelli, *Alfombras, tapices y tejidos criollos* (Buenos Aires: Guillermo Kraft, 1916).
104. Reybel, "Clemente Onelli."
105. Nari, *Políticas de maternidad*, 234.
106. Nari, *Políticas de maternidad*, 235.
107. Onelli, "'Das Ewigweiblich,'" 199–200.
108. Onelli, "'Das Ewigweiblich,'" 202.
109. Onelli, "'Das Ewigweiblich,'" 200.
110. Kristin Ruggiero, *Modernity in the Flesh: Medicine, Law, and Society in Turn-of-the-Century Argentina* (Redwood City, CA: Stanford University Press, 2004), 53.
111. Onelli, "'Das Ewigweiblich,'" 199.
112. Fernando Lahille, "Un viaje por el mundo de las garrapatas: Segunda parte," *Revista del Jardín Zoológico de Buenos Aires. Época II* 6, no. 20 (January 1910): 321–22.
113. "El zoológico social," *PBT*, 124.
114. "La artista del Zoológico," *PBT* 11, no. 481 (February 14, 1914).
115. *Guía popular . . . 1904*, 67.
116. Sanhueza Lizardi, *Un invierno en Buenos Aires*, 158.
117. Clemente Onelli, "Anécdotas de psicología zoológica (Al doctor Horacio Piñero) (Conferencia leída en la sesión ordinaria del 20 de noviembre en la Sociedad de Psicología)," *Revista del Jardín Zoológico de Buenos Aires. Época II* 6, no. 20 (January 1910): 372.
118. Clemente Onelli, "Los avestruces en libertad y en domesticidad: Sus costumbres, postura de huevos, cría de sus pollitos," *Revista del Jardín Zoológico de Buenos Aires. Época II* 1, no. 2 (July 1905): 159.
119. *Guía oficial . . . 1920–1921*, 107.
120. Hochadel, "A Global Player," 797.
121. Clemenceau, *Notas de viaje*, 40.
122. Heidrun Mühlbradt, "The Forcible Feeding of the English Suffragettes in Irish Prison: Avoiding Responsibility in Administrations," in *Apparate*, ed. Anna Echterhölter, Caspar-Fridolin Lorenz, and Tilman Richter (Berlin, Heidelberg: Springer Berlin Heidelberg, 2024), 3:122.
123. Clemenceau, *Notas de viaje*, 41.
124. Qtd. in Steven C. Hause and Anne R. Kenney, "The Limits of Suffragist Behavior: Legalism and Militancy in France, 1876–1922," *American Historical Review* 86, no. 4 (October 1981): 802.
125. In the wild, male lions may kill the cubs when they take over a pride (we all remember Scar in *The Lion King*), but Onelli does not record any aggression toward the babies on the part of the males, if it happened. Of course, this is not normal lion behavior in the wild, and the "bad mother" seems to only be reacting to the conditions she finds herself in.
126. *Guía popular . . . 1904*, 32.
127. "Una pesquisa sensacional," *Caras y Caretas* 15, no. 722 (August 3, 1912).

INTERLUDE 4

1. Pliny, *Natural History*, trans. H. Rackham, vol. III: Books 8–11, Loeb Classical Library 330 (Cambridge: Harvard University Press, 1938), 11–12.
2. *Guía popular . . . 1904*, 81.

3. Bruce Bagemihl, *Biological Exuberance: Animal Homosexuality and Natural Diversity* (New York: St. Martin's Press, 1999).
4. Bagemihl, *Biological Exuberance*, 209.
5. Bagemihl, *Biological Exuberance*, 168.
6. Bagemihl, *Biological Exuberance*, 9.
7. Bagemihl, *Biological Exuberance*, 84.
8. Paul L. Vasey and Volker Sommer, "Homosexual Behaviour in Animals: Topics, Hypotheses and Research Trajectories," in *Homosexual Behaviour in Animals: An Evolutionary Perspective*, ed. Paul L. Vasey and Volker Sommer (Cambridge: Cambridge University Press, 2006), 19–37.
9. Lahille, *Viaje al mundo*, 322.
10. Szczygielska, "Animal Sex in Public," 652–53.
11. Bagemihl, *Biological Exuberance*, 94.

CONCLUSION

1. Clemente Onelli, "El Jardín Zoológico en 1919," *Revista del Jardín Zoológico de Buenos Aires. Época II* 15, no. 60 (December 1919): 447.
2. Esparrach and Sassaroli, *Historia del Zoológico*, 74.
3. Luis Góngora, "Buenos Aires, nervio a nervio," *Mundo Argentino* 20, no. 999 (March 12, 1930).
4. David Rolla, "¿Cuándo se modernizará nuestro jardín zoológico?," *Mundo Argentino* 21, no. 1066 (June 24, 1931).
5. Esparrach and Sassaroli, *Historia del Zoológico*, 78.
6. Joseph M. Pierce, *Argentine Intimacies: Queer Kinship in an Age of Splendor, 1890–1910* (Albany: State University of New York Press, 2019), 216.
7. "Las serpientes del Jardín Zoológico," *La Prensa*, February 13, 1912.
8. "Estafa," *La Vanguardia*, October 29, 1913; A. R. R., "Jardín Zoológico de Buenos Aires," *Pacífico Magazine* 5, no. 30 (June 1915): 689.
9. Onelli, "Una mirada retrospectiva," 311.
10. Kerr, *Sex, Skulls, and Citizens*, 139.
11. Onelli, "El Jardín Zoológico en 1916," 512.
12. Onelli, "El Jardín Zoológico en 1907," 373.
13. Clemente Onelli, "El Jardín Zoológico en 1917," *Revista del Jardín Zoológico de Buenos Aires. Época II* 14, no. 53 (April 1918): 72.
14. As of 2024, the Buenos Aires and La Plata Zoos have been transformed into an ecopark and biopark, respectively. The Córdoba Zoo is now the Parque de la Biodiversidad, specializing in animal rehabilitation. In Mendoza, the zoo is undergoing renovations to become an ecopark. The Tucumán Zoo closed in the 1940s. Bahía Blanca closed to the public in 2013, with the last animals removed in 2014.
15. Clemente Onelli, "En Tucumán y en Córdoba," *Revista del Jardín Zoológico de Buenos Aires. Época II* 11, no. 44 (December 1915): 357.
16. Clemente Onelli, "Idiosincrasias individuales de los pensionistas del Jardín Zoológico," *Revista del Jardín Zoológico de Buenos Aires. Época II* 14, no. 53 (April 1918): 8.
17. Onelli, "En Tucumán," 358–59.
18. Onelli, "En Tucumán," 359.
19. Onelli, "En Tucumán," 360–61.
20. Clemente Onelli, "Los animales en la vida pública y privada (Conferencia)," *Revista del Jardín Zoológico de Buenos Aires. Época II* 11, no. 42–43 (October 1915): 176–77.

21. Onelli, "Los animales en la vida pública," 187.
22. Onelli, "En Tucumán," 361.
23. Clemente Onelli to Señor Secretario de la Intendencia Municipal, August 29, 1917. Copiador 1916–1917, 381–82, Ecoparque, Buenos Aires.

REFERENCES

ARCHIVES

Academia Nacional de la Historia de la República Argentina. Digital collections, https://repositorio.anh.org.ar.
Archivo General de la Nación. Buenos Aires, Argentina.
Biblioteca del Congreso de la Nación. Buenos Aires, Argentina.
Biblioteca Digital Agencia Española de Cooperación Internacional para el Desarrollo (AECID). Digital collection, https://bibliotecadigital.aecid.es.
Biblioteca Nacional de Maestras y Maestros. Buenos Aires, Argentina.
Biblioteca Nacional Mariano Moreno. Buenos Aires, Argentina.
Ecoparque Archive. Buenos Aires, Argentina.
Endangered Archives Programme. British Library digital collections, https://eap.bl.uk.
HathiTrust. Digital collection, https://www.hathitrust.org.
Ibero-Amerikanisches Institut. Digital collections, https://digital.iai.spk-berlin.de/viewer/index.
Instituto de Historia Argentina y Americana "Dr. Emilio Ravignani." Buenos Aires, Argentina.

BIBLIOGRAPHY

Adelman, Juliana. *Civilised by Beasts: Animals and Urban Change in Nineteenth-Century Dublin*. Manchester: Manchester University Press, 2020.
Agostoni, Claudia. *Monuments of Progress: Modernization and Public Health in Mexico City, 1876–1910*. Latin American and Caribbean Series 4. Calgary: University of Calgary Press, 2003.
Albertoni, J. B. "La Argentina país de conquista." *Estudios* 168 (June 1925): 421–37.
Allemandi, Cecilia L. "Las amas de leche y la regulación del mercado de la lactancia en la ciudad de Buenos Aires (1875–1911)." *Mora* 22 (2016): 5–24.
Altamirano, Carlos, and Beatriz Sarlo. "La Argentina del Centenario: Campo intelectual, vida literaria y temas ideológicos." In *Ensayos argentinos: De*

Sarmiento a la vanguardia, edited by Carlos Altamirano and Beatriz Sarlo, 153–90. Buenos Aires: Siglo Veintiuno Editores, 2016.

Anderson, Warwick. "Climates of Opinion: Acclimatization in Nineteenth-Century France and England." *Victorian Studies* 35, no. 2 (Winter 1992): 135–57.

Andrade, Clemente T., F. E. Giacomotti, and D. I. Espinosa. "Excursión al Jardín Zoológico." *Monitor de Educación Común* 419, no. 25 (November 30, 1907): 340–42.

Anuario Social 1913. Buenos Aires: Pablo Gadola, 1913. https://archive.org/details/anuario-social-1913/page/n93/mode/2up.

Aráoz Alfaro, Gregorio. *El libro de las madres: Manual práctico de higiene del niño, con indicaciones sobre el embarazo, parto y tratamiento de los accidentes*. Nueva edición reformada. Buenos Aires: Cabaut y Cía, 1922. https://play.google.com/books/reader?id=DpFAAQAAIAAJ&pg=GBS.PR4&hl=en.

Armus, Diego. *The Ailing City: Health, Tuberculosis, and Culture in Buenos Aires, 1870–1950*. Durham, NC: Duke University Press, 2011.

Armus, Diego, ed. *Entre médicos y curanderos: Cultura, historia y enfermedad en la América Latina moderna*. Buenos Aires: Grupo Editorial Norma, 2002.

A. R. R. "Jardín Zoológico de Buenos Aires." *Pacífico Magazine* 5, no. 30 (June 1915): 687–94.

Bagemihl, Bruce. *Biological Exuberance: Animal Homosexuality and Natural Diversity*. New York: St. Martin's Press, 1999.

Banchero, Adolfo, Coreolano Brea, and Río Raúl Franchi. "Excursión al Jardín Zoológico." *Monitor de Educación Común* 25 (1905): 250–57.

Baratay, Éric. *Animal Biographies: Toward a History of Individuals*. Translated by Lindsay Turner. Animal Voices, Animal Worlds. Athens: University of Georgia Press, 2022.

Baratay, Éric, and Elisabeth Hardouin-Fugier. *Zoo: A History of Zoological Gardens in the West*. London: Reaktion Books, 2002.

Barlaro, Pablo. "Tratado de higiene práctica elemental (1a parte)." *Revista del Jardín Zoológico de Buenos Aires. Época II* 13, no. 51 and 52 (December 1917): 305–93.

Barrancos, Dora. *La escena iluminada: Ciencias para trabajadores, 1890–1930*. Buenos Aires: Plus Ultra, 1996.

Barrington-Johnson, J. *The Zoo: The Story of London Zoo*. London: Hale, 2005.

Bates, Juandrea M. "Unaccompanied Minors and Fraudulent Fathers: Civil Law in the Unmaking of Immigrant Family in Buenos Aires, 1869–1920." *Hispanic American Historical Review* 102, no. 1 (February 1, 2022): 95–126. https://doi.org/10.1215/00182168-9497213.

Bender, Daniel E. *The Animal Game: Searching for Wildness at the American Zoo*. Cambridge, MA: Harvard University Press, 2016.

Berlant, Lauren, and Michael Warner. "Sex in Public." *Critical Inquiry* 24, no. 2 (Winter 1998): 547–66.

"Bibliografía." *La Vanguardia*, September 6, 1906.

Bignone, Héctor A. "Domingos de verano." *Caras y Caretas* 20, no. 959 (February 17, 1917).

Blasco Ibáñez, Vicente. *Argentina y sus grandezas*. Madrid: Editorial Española-Americana, 1910. https://hdl.handle.net/2027/gri.ark:/13960/t2798rb26.

Bonetti, Ángel P. *De la República Argentina y sus detractores: Homenaje a la República en su primer centenario de independencia*. Buenos Aires: El Imparcial, 1910.

Borgogno, Ariela. "*Revista Idea Latina*: Una mirada a la traducción cultural en los espacios de la diversidad." *Transfer*, no. 2 (November 2010): 47–61.

Bruce, Gary. *Through the Lion Gate: A History of the Berlin Zoo*. New York: Oxford University Press, 2017.

Bruno, Paula. "Eduardo L. Holmberg en la escena científica argentina: Ideas y acciones entre la década de 1870 y el fin-de-siglo." *Saber y Tiempo* 1, no. 1 (2015): 118–40.

Bryce, Benjamin. "Paternal Communities: Social Welfare and Immigration in Argentina, 1880–1930." *Journal of Social History* 49, no. 1 (2015): 213–36.

Bueno, Ángel. "Tipos populares." *Caras y Caretas* 16, no. 744 (January 4, 1913). https://hdl.handle.net/2027/iau.31858002550899?urlappend=%3Bseq=163%3Bownerid=13510798903556134-167.

Bunge, Carlos Octavio. *Nuestra patria: Libro de lectura para la educación nacional*. 24th ed. Buenos Aires: Estrada, 1910.

Burkhardt, Richard, Jr. "An Orangutan in Paris: Pondering Proximity at the Muséum d'histoire Naturelle in 1836." *History and Philosophy of the Life Sciences* 40, no. 1 (March 2018): 1–31.

Cándida Smith, Richard. *Improvised Continent: Pan-Americanism and Cultural Exchange*. The Arts and Intellectual Life in Modern America. Philadelphia: University of Pennsylvania Press, 2017.

Candileja, Agapito. "Interview a S.M. Cónsul III." *Caras y Caretas* 9, no. 402 (June 16, 1906). https://hdl.handle.net/2027/uva.x030598319?urlappend=%3Bseq=612%3Bownerid=27021597769316594-628.

Canton, Eliseo. *Protección a la madre y al hijo: Puericultura intra- y extrauterina*. Buenos Aires: Hunzinger, 1913. https://bipadi.ub.edu/digital/collection/salutdona/id/104996/.

Cao. "Caricaturas contemporáneas: Clemente Onelli." *Caras y Caretas* 11, no. 529 (November 21, 1908). https://hdl.handle.net/2027/uva.x004763407?urlappend=%3Bseq=674%3Bownerid=27021597769251430-726.

Cao. "La fiesta del trabajo en el zoo." *Caras y Caretas* 9, no. 396 (May 5, 1906). https://hdl.handle.net/2027/coo.31924057400388?urlappend=%3Bseq=410%3Bownerid=13510798902640092-464.

Carbone, Antonio. *Park, Tenement, Slaughterhouse: Elite Imaginaries of Buenos Aires, 1852–1880*. Frankfurt: Campus Verlag, 2022.

Cardona Leites, Amalia. "El fenómeno editorial *Caras y Caretas* y el papel del escritor en sus páginas: Horacio Quiroga y S. Fragoso Lima." *Questión: Revista Especializada en Periodismo y Comunicación* 1, no. 48 (December 2015): 78–92.
"Carnaval: El primer día de fiestas." *La Prensa*, February 19, 1917.
Carrasco, Gabriel. *Luciérnagas*. Buenos Aires: Compañía Sud-Americana de Billetes de Banco, 1905. https://hdl.handle.net/2027/uc1.b2854031.
Carreras, Sandra. "'Hay que salvar en la cuna el porvenir de la patria en peligro…': Infancia y cuestión social en Argentina (1870–1920)." In *Entre la familia, la sociedad y el Estado: Niños y jóvenes en América Latina (siglos XIX–XX)*, edited by Barbara Potthast, 143–72. Madrid: Iberoamericana, 2005.
Castellanos, Julio. "En busca de casa." *Caras y Caretas* 7, no. 358 (July 8, 1905). https://hdl.handle.net/2027/uc1.c2725150?urlappend=%3Bseq=129%3Bownerid=13510798902589243-133.
Castigat ridendo mores. "Vida social zoológica." *Revista del Jardín Zoológico de Buenos Aires. Época II* 6, no. 20 (January 1910): 373–76.
Castigat ridendo mores. "Vida social zoológica." *Revista del Jardín Zoológico de Buenos Aires. Época II* 6, no. 22 (July 1910): 141–46.
Castigat ridendo mores. "Vida social zoológica." *Revista del Jardín Zoológico de Buenos Aires. Época II* 6, no. 24 (December 1910): 290–95.
"Celebración del Día del Animal." *Billiken* 4, no. 181 (May 7, 1923): 20.
"Celebrando el año nuevo." *Caras y Caretas* 8, no. 827 (January 7, 1905). https://hdl.handle.net/2027/iau.31858002533879?urlappend=%3Bseq=156%3Bownerid=13510798903554489-188.
"Charlas del Pebete." *PBT* 5, no. 196 (August 15, 1908): 62.
Ciafardo, Eduardo. *Los niños en la ciudad de Buenos Aires (1890/1910)*. Buenos Aires: Centro Editor de América Latina, 1992.
Cid, Pío. "Los domingos del Zoo." *PBT* 5, no. 182 (May 9, 1908): 64.
Cinaghi, J. M. "Cuadros estadísticos comparativos." *Revista del Jardín Zoológico de Buenos Aires. Época II* 9, no. 34–36 (December 1913): 226–37.
Clemenceau, Georges. *Notas de viaje por la América del Sur: Argentina, Uruguay, Brasil*. Translated by Miguel Ruiz. Buenos Aires: Cabut, 1911. http://hdl.handle.net/2027/umn.31951001647997z.
"Clemente Onelli." *Mundo Israelita*, October 25, 1924.
Club de Madres. *Memoria de la 1a Semana del Nene organizada por el Club de Madres para contribuir a la disminución de la mortalidad infantil*. Buenos Aires: Wiebeck, Turtl, 1918.
Cochrane, Alasdair. "Good Work for Animals." In *Animal Labour: A New Frontier of Interspecies Justice?*, edited by Charlotte E. Blattner, Kendra Coulter, and Will Kymlicka, 48–64. Oxford: Oxford University Press, 2020.
Coitcher, Nora. "Picnics y fiestas campestres: Espacios de ocio y de derroche en la ciudad anarquista (Buenos Aires, 1897–1910)." Instituto de Arte Americano e Investigaciones Estéticas, Seminario de Crítica, no. 205. Buenos Aires, May 13, 2016. https://www.iaa.fadu.uba.ar/publicaciones/critica/0205.pdf.

Coletta, Michela. *Decadent Modernity: Civilization and "Latinidad" in Spanish America, 1880–1920*. Liverpool Latin American Studies 17. Liverpool: Liverpool University Press, 2018.

Columba, Ramón. "Apuntes del Zoológico." *Páginas de Columba* 2, no. 7 (June 1923).

Cominges, Juan de. "Palermo y el Zoológico hasta 1882." *Revista del Jardín Zoológico de Buenos Aires. Época II* 12, no. 45 (May 1916): 39–82.

"Comités Billiken." *Billiken* 6, no. 267 (December 29, 1924).

Correa Luna, Carlos. "El espíritu de rebelión." *Caras y Caretas* 7, no. 300 (July 2, 1904). https://hdl.handle.net/2027/uva.x030605268?urlappend=%3Bseq=569%3Bownerid=27021597769284367-591.

"Correo sin estampilla." *Caras y Caretas* 23, no. 1120 (March 20, 1920). https://hdl.handle.net/2027/coo.31924057355830?urlappend=%3Bseq=437%3Bownerid=13510798902640087-443.

Corsani, Patricia V. "La tarea de ornamentar plazas y paseos: El origen de las primeras esculturas del Jardín Zoológico." In *El pórtico bizantino del Jardín Zoológico de Buenos Aires: Una reflexión sobre nosotros mismos*, edited by Liliana Barela, 83–103. Buenos Aires: Dirección General Patrimonio e Instituto Histórico de la Ciudad de Buenos Aires, 2013.

Coulter, Kendra. "Toward Humane Jobs and Work-Lives for Animals." In *Animal Labour: A New Frontier of Interspecies Justice?*, edited by Charlotte E. Blattner, Kendra Coulter, and Will Kymlicka, 29–47. Oxford: Oxford University Press, 2020.

Cowie, Helen. "From the Andes to the Outback: Acclimatising Alpacas in the British Empire." *Journal of Imperial and Commonwealth History* 45, no. 4 (July 4, 2017): 551–79. https://doi.org/10.1080/03086534.2017.1353260.

"Cuadro demonstrativo de la venta de entradas al Jardín Zoológico desde el año 1914 a 1921." *Revista del Jardín Zoológico de Buenos Aires. Época II* 18, no. 66–68 (October 1922): 246.

"Cuadro demonstrativo de la venta de entradas al Jardín Zoológico durante los últimos nueve años (1901–1909)." *Revista del Jardín Zoológico de Buenos Aires. Época II* 6, no. 20 (January 1910): 382.

Currier, Charles Warren. *Lands of the Southern Cross: A Visit to South America*. Washington, DC: Spanish-American Publication Society, 1911.

Dagnino, Esteban. "En el Jardín Zoológico: Reflexiones de un visitante." *La Vanguardia*, December 28, 1906.

Dagnino, Esteban. "Parques y paseos: El Zoo." *La Vanguardia*, December 12, 1914.

Dashper, Katherine. "More-Than-Human Emotions: Multispecies Emotional Labour in the Tourism Industry." *Gender, Work & Organization* 27, no. 1 (January 2020): 24–40. https://doi.org/10.1111/gwao.12344.

Deckha, Maneesha. "Toward a Postcolonial, Posthumanist Feminist Theory: Centralizing Race and Culture in Feminist Work on Nonhuman Animals." *Hypatia* 27, no. 3 (Summer 2012): 527–45.

de Courcy, Catherine. *Dublin Zoo: An Illustrated History*. Cork: Collins, 2009.

Delaney, Jean. "Imagining 'El Ser Argentino': Cultural Nationalism and Romantic Concepts of Nationhood in Early Twentieth-Century Argentina." *Journal of Latin American Studies* 34, no. 3 (August 2002): 625–58.

"De las provincias." *La Vanguardia*, October 23, 1918.

del Pino, Diego A. *Clemente Onelli: El más criollo de los tanos*. Buenos Aires: Ediciones Turísticas, 2005.

del Pino, Diego A. *Historia del Jardín Zoológico Municipal*. Cuadernos de Buenos Aires 55. Buenos Aires: Cuadernos de Buenos Aires, 1979.

Departamento Nacional del Trabajo. *Boletín No. 42 del Departamento Nacional del Trabajo: Anuario estadístico 1917*. Buenos Aires: Gotelli y Cía., 1919. https://www.argentina.gob.ar/sites/default/files/mintrabajo/1907-1921%20boletin%20trimestral/mteyss-biblioteca-dnt-trimestal-1919_42.pdf.

"Desde los dominios de Onelli: Vuelve a escribirnos un tigre de Bengala." *La Razón*, September 4, 1922.

Devoto, Fernando. *Historia de la inmigración en la Argentina*. 2nd ed. Buenos Aires: Sudamericana, 2004.

Díaz, Marcela Liliana, and Cristina Fernández. *Guía del patrimonio cultural del Jardín Zoológico de la Ciudad de Buenos Aires*. Buenos Aires: DyPContenidos, 2016. https://www.iaa.fadu.uba.ar/publicaciones/otros/GuiaPatrimonioCulturalZOOBA.pdf.

Di Liscia, María Silvia. "Marcados en la piel: Vacunación y viruela en Argentina (1870–1910)." *Ciencia & Saúde Coletiva* 16, no. 2 (2011): 409–22. https://doi.org/10.1590/S1413-81232011000200005.

Di Liscia, María Silvia, and Ernesto Lázaro Bohoslavsky. *Instituciones y formas de control social en América Latina, 1840–1940: Una revisión*. Buenos Aires: Prometeo, 2005.

Di Liscia, María Silvia, and Graciela Nélida Salto, eds. *Higienismo, educación y discurso en la Argentina, 1870–1940*. Santa Rosa, La Pampa: Editorial de la Universidad Nacional de La Pampa, 2004.

Dovio, Mariana. "El Instituto de Criminología y la 'mala vida' entre 1907 y 1913." *Anuario de la Escuela de Historia Virtual* 4, no. 4 (2013): 93–117.

Durán, Manuel. "Sexualidad, producción y trabajo en el discurso higienista y eugenésico en Chile y Argentina, 1860–1930." *Revista Nomadías* 23 (July 2017): 31–52.

Dutton, Christopher L., Amanda L. Subalusky, Alvaro Sanchez, Sylvie Estrela, Nanxi Lu, Stephen K. Hamilton, Laban Njoroge, Emma J. Rosi, and David M. Post. "The Meta-Gut: Community Coalescence of Animal Gut and Environmental Microbiomes." *Scientific Reports* 11, no. 1 (November 30, 2021). https://doi.org/10.1038/s41598-021-02349-1.

Edwards, Erika Denise. *Hiding in Plain Sight: Black Women, the Law, and the Making of a White Argentine Republic*. Tuscaloosa: University of Alabama Press, 2021.

"El acontecimiento zoo-cial de la semana." *Caras y Caretas* 9, no. 404 (June 30, 1906). https://hdl.handle.net/2027/uva.x030598319?urlappend=%3Bseq=799%3Bownerid=27021597769305240-947.
"El carnaval de los niños." *Plus Ultra* 2, no. 11 (March 1917).
El del verde gabán. "Visita de duelo." *PBT* 5, no. 196 (August 15, 1908): 64.
"El Jardín Zoológico de duelo: Muerte de Petronio." *La Prensa*, January 7, 1908.
El Judío Errante. "Baedeker Humorístico de la República Argentina: El Botánico y el Zoológico." *PBT* 10, no. 453 (August 2, 1913).
"El Parque Zoológico del Sur." *PBT* 11, no. 517 (October 24, 1914).
"El pequeño dromedario del Zoológico." *PBT* 46, no. 360 (October 21, 1911).
"El Zoológico no progresa: Carta de un león." *La Razón*, August 29, 1922.
"El zoológico social." *PBT* 5, no. 201 (September 19, 1908): 124–25.
"En el ferrocarril del Zoo." *Sherlock Holmes* 1, no. 26 (December 26, 1911): 50.
"En el Jardín Zoológico: Búfalo enfurecido." *La Prensa*, October 22, 1911.
"En el Jardín Zoológico: Protestas infantiles." *La Prensa*, May 20, 1917.
"En el Palacio Novedades: El mono Cónsul, patinador, sportsman." *Caras y Caretas* 13, no. 613 (July 2, 1910). https://hdl.handle.net/2027/iau.31858002534109?urlappend=%3Bseq=140%3Bownerid=13510798903553927-184.
"En el Zoo." *La Nación*, August 16, 1905.
"En el Zoo." *La Nación*, June 25, 1914.
"En el Zoológico." *PBT* 10, no. 440 (May 3, 1913). https://digital.iai.spk-berlin.de/viewer/image/86151033X/77/.
Esparrach, Eduardo, and Juan Carlos Sassaroli. *Historia del Zoológico de Buenos Aires: 1888–2020*. Buenos Aires: Vázquez Mazzini Editores, 2022.
"Estafa." *La Vanguardia*, October 29, 1913.
Estrada, Santiago. *Viajes y otras páginas literarias*, 4th ed. Edited by Ricardo Ryan. Buenos Aires: Ángel Estrada, 1942.
Eujanian, Alejandro C. *Historia de revistas argentinas, 1900–1950: La conquista del pueblo*. Buenos Aires: Asociación Argentina de Editores de Revistas, 1999.
Everest, Sophie. "'Under the Skin': The Biography of a Manchester Mandrill." In *The Afterlives of Animals: A Museum Menagerie*, edited by J. M. M. Alberti, 75–91. Charlottesville: University of Virginia Press, 2011.
Few, Martha, and Zeb Tortorici, eds. *Centering Animals in Latin American History*. Durham, NC: Duke University Press, 2013.
Flack, Andrew. *The Wild Within: Histories of a Landmark British Zoo*. Charlottesville: University of Virginia Press, 2018.
Fierro, Martín. "Música de actualidad." *Ideas y figuras* 1, no. 1 (May 13, 1909).
Fierro, Martín, hijo. "Jardín Zoológico (Palermo)." *Iris* 1, no. 17 (November 23, 1899).
Fudge, Erica. *Brutal Reasoning: Animals, Rationality, and Humanity in Early Modern England*. Ithaca, NY: Cornell University Press, 2006.
Gallardo, Ángel. "Las excepciones al Mendelismo: Una carta del Dr. A Gallardo." *Revista del Jardín Zoológico de Buenos Aires. Época II* 4, no. 15 (September 1908): 305–6.

Gálvez, Manuel. *El solar de la raza*. 5th ed. Madrid: Saturnino Calleja, 1920. https://archive.org/details/elsolardelarazaooglve.

Gálvez, Manuel. *Nacha Regules*. Edited by Aníbal Jarkowski. Buenos Aires: Eterna Cadencia, 2010.

Giorgi, Gabriel. *Formas comunes: Animalidad, cultura, biopolítica*. Buenos Aires: Eterna Cadencia, 2014.

Giudice, Aldo Mario, Eduardo Esparrach, and María Teresa Ferrero de Roqué. "Serie: Elefantes rioplatenses y zoológicos: La inecuación perfecta. (III) Elefantes en el Zoológico de Buenos Aires." *Revista Boletín Biológica*, no. 39, año 12 (2018): 11–18.

Gomi, Tarō. *Everyone Poops*. Translated by Amanda Mayer Stinchecum. La Jolla: Turtleback Books, 2018.

Góngora, Luis. "Buenos Aires, nervio a nervio" *Mundo Argentino* 20, no. 999 (March 12, 1930).

González Herrera, M. "Causas de muerte en los Jardines Zoológicos (continuación)." *Revista del Jardín Zoológico de Buenos Aires. Época II* 1, no. 4 (December 1905): 343–47.

Gordillo, Gastón, and Silvia Hirsch. "Indigenous Struggles and Contested Identities in Argentina: Histories of Invisibilization and Reemergence." *Journal of Latin American Anthropology* 8, no. 3 (2003): 4–30.

Gorelik, Adrián. *The Grid and the Park: Public Space and Urban Culture in Buenos Aires, 1887–1936*. Translated by Natalia Majluf. Pittsburgh, PA: Latin America Research Commons, 2022.

Goyo Cuello. "Equitación infantil en el Zoo." *Caras y Caretas* 22, no. 1075 (May 10, 1910). https://hdl.handle.net/2027/uc1.c2725292?urlappend=%3Bseq=159%3Bownerid=13510798902586574-165.

Grazian, David. *American Zoo: A Sociological Safari*. Princeton, NJ: Princeton University Press, 2015.

Grosse, Pascal. "Turning Native?: Anthropology, German Colonialism, and the Paradoxes of the 'Acclimatization Question,' 1885–1914." In *Worldly Provincialism: German Anthropology in the Age of Empire*, edited by H. Glenn Penny and Matti Bunzi, 179–97. Ann Arbor: University of Michigan Press, 2003.

Guía oficial ilustrada del Jardín Zoológico Municipal de Buenos Aires 1908–1909. Buenos Aires: Gaibisso, 1908.

Guía oficial ilustrada del Jardín Zoológico Municipal de Buenos Aires 1920–1921. Buenos Aires: Compañía General de Fósforos, 1920.

Guía oficial municipal del Jardín Zoológico 1906–1907. Buenos Aires: Vasco Fogli, 1906.

Guía popular del Jardín Zoológico Municipal de Buenos Aires 1904. Buenos Aires: A. Rothkopf, 1904.

Gutiérrez, Leandro H., and Juan Suriano. "Workers' Housing and Living Conditions in Buenos Aires, 1880–1930." In *Essays in Argentine Labor*

History, 1870–1930, edited by Jeremy Adelman, 35–51. Basingstoke: Macmillan, 1992.

Guy, Donna J. *Sex and Danger in Buenos Aires: Prostitution, Family, and Nation in Argentina*. Engendering Latin America. Lincoln: University of Nebraska Press, 1991.

"Hablando con Onelli." *PBT* 14, no. 640 (March 3, 1917). https://digital.iai.spk-berlin.de/viewer/image/861636589/38/.

Hammerton, John Alexander. *The Real Argentine: Notes and Impressions of a Year in the Argentine and Uruguay*. New York: Dodd, Mead, 1915.

Hanson, Elizabeth. *Animal Attractions: Nature on Display in American Zoos*. Princeton, NJ: Princeton University Press, 2002.

Hause, Steven C., and Anne R. Kenney. "The Limits of Suffragist Behavior: Legalism and Militancy in France, 1876–1922." *American Historical Review* 86, no. 4 (October 1981): 781–806.

"Hazañas trágicas y ridículas de los pensionistas del Zoo." *Caras y Caretas* 9, no. 413 (September 1, 1906). https://hdl.handle.net/2027/hvd.32044094187200?urlappend=%3Bseq=828%3Bownerid=27021597764401591-838.

Hochadel, Oliver. "A Global Player from the South: The Jardín Zoológico de Buenos Aires and the Transnational Network of Zoos in the Early Twentieth Century." *História, Ciências, Saúde-Manguinhos* 29, no. 3 (2002): 789–812. https://doi.org/10.1590/S0104-59702022000300012.

Hochadel, Oliver. "Science at the Zoo: An Introduction." *Centaurus* 64, no. 3 (October 2022): 561–90. https://doi.org/10.1484/J.CNT.5.132186.

Hochman, Gilberto. *The Sanitation of Brazil: Nation, State, and Public Health, 1889–1930*. Translated by Diane R. Grosklaus Whitty. Urbana: University of Illinois Press, 2016.

Holmberg, Eduardo L. "A la prensa." *Revista del Jardín Zoológico de Buenos Aires* 1, no. 1 (January 15, 1893): 3.

Holmberg, Eduardo L. "El Jardín Zoológico en 1890: La primera guía del Jardín Zoológico de Buenos Aires." *Revista del Jardín Zoológico de Buenos Aires* 1, no. 8 (August 15, 1893): 225–54.

Holmberg, Eduardo L. "El reglamento del Jardín Zoológico." *Revista del Jardín Zoológico de Buenos Aires* 1, no. 1 (January 15, 1893): 5–19.

Holmberg, Eduardo L. "Informe anual del Director del Jardín Zoológico correspondiente al año 1893." *Revista del Jardín Zoológico de Buenos Aires* 2, no. 1 (January 15, 1894): 3–32.

Holmberg, Eduardo L. "Jardín Zoológico de Buenos Ayres: Mamíferos. Cueros armados existentes en el museo del Jardín el día 31 de diciembre de 1892." *Revista del Jardín Zoológico de Buenos Aires* 1, no. 4 (April 15, 1893): 116–19.

Holmberg, Eduardo L. "Neán: La elefante del Jardín Zoológico." *Revista del Jardín Zoológico de Buenos Aires* 1, no. 3 (March 15, 1893): 65–75.

Holmberg, Eduardo L. "Osteomalacia." *Revista del Jardín Zoológico de Buenos Aires* 1, no. 2 (February 15, 1893): 37–39.

Horta Duarte, Regina. "'El zoológico del porvenir': Narrativas y memorias de nación sobre el Zoológico de Chapultepec, Ciudad de México, siglo XX." *Historia Crítica* 72 (2019): 93–113.

Horta Duarte, Regina. "Zoos in Latin America." *Oxford Research Encyclopedia of Latin American History*. September 26, 2017. https://doi.org/10.1093/acrefore/9780199366439.013.439.

Hribal, Jason. "Animals, Agency, and Class: Writing the History of Animals from Below." *Human Ecology Review* 14, no. 1 (Summer 2007): 101–12.

Hribal, Jason. *Fear of the Animal Planet: The Hidden History of Animal Resistance*. Edinburgh: AK, 2010.

Huber, Augusto. "El karakul." *Revista del Jardín Zoológico de Buenos Aires. Época II* 8, no. 29 (April 1912): 53–59.

Huergo, Juan Carlos. "Un domingo en el Zoo." *La Nación*, August 14, 1921.

Hui, Alexandra, Lissa Roberts, and Seth Rockman. "Introduction: Launching a Labor History of Science." *Isis* 114, no. 4 (December 2023): 817–26. https://doi.org/10.1086/727646.

Humo, Marcus. "Cosas del año nuevo." *Caras y Caretas* 11, no. 484 (January 11, 1908). https://hdl.handle.net/2027/uva.x030598326?urlappend=%3Bseq=271%3Bownerid=27021597769308243-287.

"Instrucción popular: Sociedad Luz, la excursión al Jardín Zoológico." *La Vanguardia*, October 21, 1913.

Ito, Takashi. "History of the Zoo." In *Handbook of Historical Animal Studies*, edited by Mieke Roscher, André Krebber, and Brett Mizelle, 439–55. Berlin: De Gruyter Oldenbourg, 2021.

Ito, Takashi. *London Zoo and the Victorians, 1828–1859*. Royal Historical Society Studies in History. New Series. Woodbridge, Suffolk: Royal Historical Society, 2014.

Jakob, Christofredo. "Problemas actuales de psiquiatría general y sus relaciones con las ciencias sociales y jurídicas." *Revista de Criminología, Psiquiatría y Medicina Legal* 4 (1917): 3–45.

Jakob, Christofredo. "Tratado de biología general y especial: Biología vegeto-animal elemental." *Revista del Jardín Zoológico de Buenos Aires. Época II* 12, no. 48 (December 1916): 529–693.

Jakob, Christofredo. "Tratado de biología general y especial para el uso de la enseñanza elemental, secundaria y superior en la República Argentina." *Revista del Jardín Zoológico de Buenos Aires. Época II* 11, no. 42–43 (October 1915): 274–327.

Jakob, Christofredo, and Clemente Onelli. *Atlas del cerebro de los mamíferos de la República Argentina: Estudios anatómicos, histológicos y biológicos comparados, sobre la evolución de los hemisferios y de la corteza cerebral.*

Buenos Aires: Guillermo Kraft, 1913. https://hdl.handle.net/2027/njp.32101076140043.
"Jardín Zoológico." *Caras y Caretas* 7, no. 280 (February 13, 1904).
"Jardín Zoológico: El día de un director." *Caras y Caretas* 15, no. 709 (May 4, 1912). https://hdl.handle.net/2027/uva.x030598347?urlappend=%3Bseq=655%3Bownerid=27021597769317954-653.
Kerr, Ashley Elizabeth. *Sex, Skulls, and Citizens: Gender and Racial Science in Argentina (1860–1910)*. Nashville, TN: Vanderbilt University Press, 2020.
Kisling, Vernon N., ed. *Zoo and Aquarium History: Ancient Animal Collections to Conservation Centers*. Second edition. New York: CRC Press, 2022.
Kohl, Alejandro. *Higienismo argentino: Historia de una utopía. La salud en el imaginario colectivo de una época*. Buenos Aires: Editorial Dunken, 2006.
"La artista del Zoológico." *PBT* 11, no. 481 (February 14, 1914).
"La comida de las fieras." *Fray Mocho* 2, no. 7 (October 3, 1913).
"La girafa 'Mimí.'" *Caras y Caretas* 15, no. 709 (May 4, 1912). https://hdl.handle.net/2027/uva.x030598347?urlappend=%3Bseq=658%3Bownerid=27021597769307731-666.
"La higiene en el Zoo." *Billiken* 3, no. 143 (August 14, 1922).
Lahille, Fernando. "La zoología, los conocimientos que necesita y los estudios que abarca." *Revista del Jardín Zoológico de Buenos Aires. Época II* 3, no. 10 (July 1907): 111–31.
Lahille, Fernando. "Un viaje por el mundo de las garrapatas: Segunda parte." *Revista del Jardín Zoológico de Buenos Aires. Época II* 6, no. 20 (January 1910): 287–341.
"La inauguración de la Cabrería Muncipal." *Caras y Caretas* 15, no. 734 (October 26, 1912). https://hdl.handle.net/2027/iau.31858002550873?urlappend=%3Bseq=1173%3Bownerid=13510798903554933-1219.
"La jirafa del Zoo." *La Semana Universal* 1 (May 2, 1912). https://hdl.handle.net/2027/txu.059173018059020?urlappend=%3Bseq=140%3Bownerid=27021597767411113-146.
"La muerte de Petronio." *La Nación*, January 7, 1908.
Lanús, Adolfo. "Impresiones del Zoológico: El desayuno." *Caras y Caretas* 22, no. 1103 (November 22, 1919). https://hdl.handle.net/2027/coo.31924057355673?urlappend=%3Bseq=1064%3Bownerid=13510798902637577-1098.
"Las serpientes del Jardín Zoológico." *La Prensa*, February 13, 1912.
Laveaga, Gabriela Soto. "Worker Once Known: Thinking with Disposable, Discarded, Mislabeled, and Precariously Employed Laborers in History of Science." *Isis* 114, no. 4 (December 2023): 834–40. https://doi.org/10.1086/727760.
"La visita de Mr. Roosevelt." *Caras y Caretas* 16, no. 789 (November 15, 1913). https://hdl.handle.net/2027/iau.31858002550949?urlappend=%3Bseq=355%3Bownerid=13510798903554131-401.

Lavrin, Asunción. *Women, Feminism, and Social Change in Argentina, Chile, and Uruguay, 1890–1940*. Lincoln: University of Nebraska Press, 1995.

Lehmann-Nitsche, Robert. "Aclimatación de la perdiz grande y de la martineta en Alemania." *Revista del Jardín Zoológico de Buenos Aires. Época II* 3, no. 10 (July 1907): 132–37.

Libedinsky, Marta Alicia. "La excursión escolar en la Ciudad de Buenos Aires (fines del siglo XIX y segunda década del siglo XX): Diálogos histórico-didácticos entre pasado, presente y viceversa." PhD diss., Universidad Nacional del Centro de la Provincia de Buenos Aires, 2021.

"Liste supplémentaire des membres de la société." *Bulletin de la Societé Nationale de Acclimatation de France* 58 (January 1, 1911).

Loisel, Gustave. *Histoire des ménageries de l'antiquité à nos jours*. Vol. 3. Paris: Octave Doin et Fils, 1912.

Lombroso, Cesare. "Una opinión honrosa." *Revista del Jardín Zoológico de Buenos Aires. Época II* 4, no. 17 (April 1909): 79.

Losada, Leandro. *La alta sociedad en la Buenos Aires de la belle epoque: Sociabilidad, estilo de vida e identidades*. Buenos Aires: Siglo XXI, 2008.

"Los peones del Jardín Zoológico." *La Nación*, October 22, 1918.

Lugones, Leopoldo. *Didáctica*. Buenos Aires: Otero, 1910. https://hdl.handle.net/2027/njp.32101068146362.

Lynch, Ricardo. "L'alimentation et l'instinct de l'alimentation chez l'homme et chez les animaux." *Revista del Jardín Zoológico de Buenos Aires. Época II* 7, no. 25 (April 1911): 15–53.

"Majaderías." *La Vanguardia*, September 8, 1908.

Malamud, Randy. *Reading Zoos: Representations of Animals and Captivity*. New York: New York University Press, 1998.

Marthur, Eduardo. "Pic-nic popular." *Caras y Caretas* 17, no. 811 (April 18, 1914). https://hdl.handle.net/2027/iau.31858002550964?urlappend=%3Bseq=853%3Bownerid=13510798903554322-867.

Martín, Ana Laura. "Partear y cuidar en Buenos Aires (1877–1920): Una aproximación comparativa." *Anuario del Instituto de Historia Argentina* 18, no. 1 (June 2018).

Marvin, Garry. "Wolves in Sheep's (and Others') Clothing." In *Beastly Natures: Animals, Humans, and the Study of History*, edited by Dorothee Brantz, 59–78. Charlottesville: University of Virginia Press, 2010.

McDonald, Tracy. "Sculpting Dinah with the Blunt Tools of the Historian." In *Zoo Studies: A New Humanities*, edited by Tracy McDonald and Daniel Vandersommers, 93–118. Montreal: McGill-Queen's University Press, 2019.

McDonald, Tracy, and Daniel Vandersommers, eds. *Zoo Studies: A New Humanities*. Montreal: McGill-Queen's University Press, 2019.

McLeod, Katherine. "The Call of the Hoatzin: Ecology, Evolution, and Eugenics at the Bronx Zoo." *Centaurus* 64, no. 3 (October 2022): 683–704. https://doi.org/10.1484/J.CNT.5.132105.

Méndez, Patricia, and Rodrigo Gutiérrez Viñuales. "Buenos Aires en el Centenario: Edificación de la nación y la nación edificada." *Apuntes* 19, no. 2 (2007): 216–27.
"Menudencias." *Caras y Caretas* 9, no. 387 (March 3, 1906). https://hdl.handle.net/2027/iau.31858002533911?urlappend=%3Bseq=928%3Bownerid=13510798903554928-934.
"Menudencias," *Caras y Caretas* 9, no. 401 (June 8, 1906), https://hdl.handle.net/2027/uva.x030598319?urlappend=%3Bseq=530%3Bownerid=27021597769316594-558.
"Menudencias." *Caras y Caretas* 15, no. 709 (May 4, 1912). https://hdl.handle.net/2027/uva.x030598347?urlappend=%3Bseq=662%3Bownerid=27021597769317954-660.
Miatello, Hugo. "Una institución benéfica: La Universidad Popular de La Boca." *Caras y Caretas* 22, no. 1087 (August 2, 1919). https://hdl.handle.net/2027/iau.31858002547762?urlappend=%3Bseq=568%3Bownerid=13510798903550090-606.
Miller, Ian Jared. *The Nature of the Beasts: Empire and Exhibition at the Tokyo Imperial Zoo*. Berkeley: University of California Press, 2013.
Miller, Nicola. *In the Shadow of the State: Intellectuals and the Quest for National Identity in Twentieth-Century Spanish America*. Critical Studies in Latin American and Iberian Cultures. London; New York: Verso, 1999.
Minard, Peter Maxwell. *All Things Harmless, Useful, and Ornamental: Environmental Transformation through Species Acclimatization, from Colonial Australia to the World*. Flows, Migrations, and Exchanges. Chapel Hill: University of North Carolina Press, 2019.
Miranda, Marisa. "Maternidad y biopolítica en la Argentina: Gregorio Aráoz Alfaro, *El libro de las madres* y la eugenesia (1870–1955)." *Passagens: Revista Internacional de História Política e Cultura Jurídica* 11, no. 2 (August 2019): 156–76.
Miranda, Marisa. "Matrimonio y procreación en la ortodoxia eugénica argentina." *Sociohistórica* 17–18 (first and second semester 2005): 151–78.
Montagne, Edmundo. "El viaje de 'Mundial.'" *Mundial Magazine* 2, no. 19 (November 1912): 589–609.
Morin, Karen M. *Carceral Space, Prisoners and Animals*. Routledge Human-Animal Studies. Abingdon: Routledge, 2018.
Mortimer-Sandilands, Catriona. "Unnatural Passions?: Notes Toward a Queer Ecology." *Invisible Culture* 9 (2005). https://www.rochester.edu/in_visible_culture/Issue_9/title9.html.
"Mucanga burguesa." *La Vanguardia*, February 13, 1904.
Mühlbradt, Heidrun. "The Forcible Feeding of the English Suffragettes in Irish Prison: Avoiding Responsibility in Administrations." In *Apparate*, edited by Anna Echterhölter, Caspar-Fridolin Lorenz, and Tilman Richter, 3:117–32. AdminiStudies: Formen und Medien der Verwaltung. Berlin, Heidelberg:

Springer Berlin Heidelberg, 2024. https://doi.org/10.1007/978-3-662-67712-4_7.
Mullan, Bob, and Garry Marvin. *Zoo Culture*, 2nd ed. Urbana: University of Illinois Press, 1999.
Munck, Ronaldo. *Argentina: From Anarchism to Peronism*. London: Zed Books, 1987.
Nari, Marcela María Alejandra. "El feminismo frente a la cuestión de la mujer en las primeras décadas del siglo XX." In *La cuestión social en la Argentina, 1870–1943*, edited by Juan Suriano, 277–99. Buenos Aires: La Colmena, 2000.
Nari, Marcela. *Políticas de maternidad y maternalismo político: Buenos Aires, 1890–1940*. Buenos Aires: Biblos, 2004.
"Novedades del Zoo." *Sherlock Holmes* 1, no. 16 (October 17, 1911): 5.
Novillo Quiroga, Diego. *Sabañones*. Buenos Aires: Tixi & Schaffner, 1922.
Obregón. "Congreso de Animales: Sesión extraordinaria en honor del Doctor Albarracín." *El Gladiador* 3, no. 140 (August 5, 1904).
Onelli, Clemente. "A la par con el yankee." *El Monitor de Educación Común* 36, no. 540 (December 31, 1917): 133–35.
Onelli, Clemente. *Alfombras, tapices y tejidos criollos*. Buenos Aires: Guillermo Kraft, 1916.
Onelli, Clemente. "Algunas correspondencias cangeadas en el trimestre." *Revista del Jardín Zoológico de Buenos Aires. Época II* 4, no. 13 (April 1908): 100–105.
Onelli, Clemente. "Ameghino en la ciencia y en la vida privada." *Revista del Jardín Zoológico de Buenos Aires. Época II* 11, no. 42–43 (October 1915): 217–33.
Onelli, Clemente. "Anécdotas de psicología zoológica (Al doctor Horacio Piñero) (Conferencia leída en la sesión ordinaria del 20 de noviembre en la Sociedad de Psicología)." *Revista del Jardín Zoológico de Buenos Aires. Época II* 6, no. 20 (January 1910): 345–72.
Onelli, Clemente. "Apuntes sobre el Jardín Zoológico para el Dr. Gustavo Loisel del Laboratorio de Embriología de París." *Revista del Jardín Zoológico de Buenos Aires. Época II* 3, no. 11 (October 1907): 277–88.
Onelli, Clemente. "Aumento vegetativo: Los principales productos del año." *Revista del Jardín Zoológico de Buenos Aires. Época II* 3, no. 9 (April 1907): 68–70.
Onelli, Clemente. "Buen trato hacia los animales, enfermedades que contagian, etc.: Conferencia tenida por el Director del Jardín Zoológico a los miembros de la Universidad Popular 'Luz.'" *Revista del Jardín Zoológico de Buenos Aires. Época II* 2, no. 7 (October 1906): 231–40.
Onelli, Clemente. "Cómo se curan los ignorantes: Remedios empíricos sacados de los animales." *Revista del Jardín Zoológico de Buenos Aires. Época II* 1, no. 2 (July 1905): 133–40.
Onelli, Clemente. "Cómo se pierde y cómo se hace perder el tiempo." *Revista del Jardín Zoológico de Buenos Aires. Época II* 7, no. 25 (April 1911): 54–87.

Onelli, Clemente. "Correspondencia." *Revista del Jardín Zoológico de Buenos Aires. Época II* 4, no. 14 (July 1908): 202–5.

Onelli, Clemente. "'Das Ewigweiblich' o La esfinge sin enigmas." *Revista del Jardín Zoológico de Buenos Aires. Época II* 10, no. 39–40 (December 1914): 191–206.

Onelli, Clemente. "Desencantos e ilusiones de un naturalista (Conferencia leída en la Sociedad Luz)." *Revista del Jardín Zoológico de Buenos Aires. Época II* 12, no. 48 (December 1916): 487–506.

Onelli, Clemente. "El afecto paternal y el de la madre." *Revista del Jardín Zoológico de Buenos Aires. Época II* 3, no. 9 (April 1907): 52–59.

Onelli, Clemente. "El avestruz criollo en moda." *Anales de la Sociedad Rural Argentina*, no. 42–43 (1908): 70–73.

Onelli, Clemente. "El empleado en la colmena humana: Conferencia leída la noche del 23 de septiembre de 1916 en el Club Sindical de Empleados." *Revista del Jardín Zoológico de Buenos Aires. Época II* 12, no. 47 (October 1916): 329–44.

Onelli, Clemente. "El Jardín Zoológico en 1907: Vistas para el futuro." *Revista del Jardín Zoológico de Buenos Aires. Época II* 3, no. 12 (December 1907): 363–75.

Onelli, Clemente. "El Jardín Zoológico en 1909." *Revista del Jardín Zoológico de Buenos Aires. Época II* 6, no. 21 (April 1910): 46–55.

Onelli, Clemente. "El Jardín Zoológico en 1910." *Revista del Jardín Zoológico de Buenos Aires. Época II* 6, no. 24 (December 1910): 298–304.

Onelli, Clemente. "El Jardín Zoológico en 1916." *Revista del Jardín Zoológico de Buenos Aires. Época II* 12, no. 48 (December 1916): 507–13.

Onelli, Clemente. "El Jardín Zoológico en 1917." *Revista del Jardín Zoológico de Buenos Aires. Época II* 14, no. 53 (April 1918): 70–74.

Onelli, Clemente. "El Jardín Zoológico en 1919." *Revista del Jardín Zoológico de Buenos Aires. Época II* 15, no. 60 (December 1919): 447–49.

Onelli, Clemente. "El Jardín Zoológico en el año 1908 y 1909." *Revista del Jardín Zoológico de Buenos Aires. Época II* 4, no. 16 (December 1908): 395–400.

Onelli, Clemente. "El Jardín Zoológico en el extranjero." *Revista del Jardín Zoológico de Buenos Aires. Época II* 5, no. 19 (October 1909): 270–71.

Onelli, Clemente. "El Jardín Zoológico y sus anexos en 1915." *Revista del Jardín Zoológico de Buenos Aires. Época II* 11, no. 44 (December 1915): 419–37.

Onelli, Clemente. "El karakul criollo." *Revista del Jardín Zoológico de Buenos Aires. Época II* 8, no. 30 (July 1912): 123–28.

Onelli, Clemente. "El Parque Zoológico del Sud." *Revista del Jardín Zoológico de Buenos Aires. Época II* 10, no. 39–40 (December 1914): 139–50.

Onelli, Clemente. "El peligro de las moscas." *Revista del Jardín Zoológico de Buenos Aires. Época II* 6, no. 24 (December 1910): 281–82.

Onelli, Clemente. "El sionismo integral." *Vida Nuestra* 1, no. 12 (June 1918): 286–87.

Onelli, Clemente. "En Tucumán y en Córdoba." *Revista del Jardín Zoológico de Buenos Aires. Época II* 11, no. 44 (December 1915): 357–61.

Onelli, Clemente. "Escuela municipal de telares domésticos." *Revista del Jardín Zoológico de Buenos Aires. Época II* 18, no. 66–68 (October 1922): 163–66.

Onelli, Clemente. "Exámenes coprológicos en los animales." *Revista del Jardín Zoológico de Buenos Aires. Época II* 2, no. 6 (July 1906): 117–24.

Onelli, Clemente. "Foseta del vermis en los cráneos de indios, mamíferos y saurianos." *Revista del Jardín Zoológico de Buenos Aires. Época II* 1, no. 4 (December 1905): 352–56.

Onelli, Clemente. "Ganadería de pelo fino." *Anales de la Sociedad Rural Argentina* 42–43 (1908): 55–62.

Onelli, Clemente. "Ganadería exótica." *Anales de la Sociedad Rural Argentina* 40 (May–June 1906): 65–69. https://hdl.handle.net/2027/uc1.c2635064?urlappend=%3Bseq=304%3Bownerid=13510798902947619-344.

Onelli, Clemente. "Gorriones sovietistas." *Fray Mocho* 17, no. 828 (March 6, 1928): 38.

Onelli, Clemente. "Ideas disonantes sobre educandos y educacionistas." *Revista del Jardín Zoológico de Buenos Aires. Época II* 9, no. 34–36 (December 1913): 173–90.

Onelli, Clemente. "Idiosincrasias individuales de los pensionistas del Jardín." *Revista del Jardín Zoológico de Buenos Aires. Época II* 1, no. 3 (October 1905): 221–36.

Onelli, Clemente. "Idiosincrasias individuales de los pensionistas del Jardín." *Revista del Jardín Zoológico de Buenos Aires. Época II* 2, no. 5 (April 1906): 1–14.

Onelli, Clemente. "Idiosincrasias individuales de los pensionistas del Jardín." *Revista del Jardín Zoológico de Buenos Aires. Época II* 2, no. 6 (July 1906): 81–100.

Onelli, Clemente. "Idiosincrasias de los pensionistas del Jardín." *Revista del Jardín Zoológico de Buenos Aires. Época II* 2, no. 7 (October 1906): 171–86.

Onelli, Clemente. "Idiosincrasias individuales de los pensionistas del Jardín." *Revista del Jardín Zoológico de Buenos Aires. Época II* 3, no. 9 (April 1907): 1–18.

Onelli, Clemente. "Idiosincrasias individuales de los pensionistas del Jardín Zoológico." *Revista del Jardín Zoológico de Buenos Aires. Época II* 3, no. 11 (October 1907): 195–211.

Onelli, Clemente. "Idiosincrasias individuales de los pensionistas del Jardín Zoológico." *Revista del Jardín Zoológico de Buenos Aires. Época II* 3, no. 12 (December 1907): 299–317.

Onelli, Clemente. "Idiosincrasias individuales de los pensionistas del Jardín." *Revista del Jardín Zoológico de Buenos Aires. Época II* 4, no. 13 (April 1908): 1–20.

Onelli, Clemente. "Idiosincrasias individuales de los pensionistas del Jardín Zoológico." *Revista del Jardín Zoológico de Buenos Aires. Época II* 4, no. 15 (September 1908): 213–26.

Onelli, Clemente. "Idiosincrasias individuales de los pensionistas del Jardín Zoológico." *Revista del Jardín Zoológico de Buenos Aires. Época II* 4, no. 16 (December 1908): 327–45.

Onelli, Clemente. "Idiosincrasias individuales de los pensionistas del Jardín Zoológico." *Revista del Jardín Zoológico de Buenos Aires. Época II* 5 [identified mistakenly as 6 in the original printing of the journal], no. 18 (July 1909): 81–99.

Onelli, Clemente. "Idiosincrasias individuales de los pensionistas del Jardín Zoológico." *Revista del Jardín Zoológico de Buenos Aires. Época II* 5, no. 19 (October 1909): 179–202.

Onelli, Clemente. "Idiosincrasias individuales de los pensionistas del Jardín Zoológico." *Revista del Jardín Zoológico de Buenos Aires. Época II* 6, no. 21 (April 1910): 3–13.

Onelli, Clemente. "Idiosincrasias individuales de los pensionistas del Jardín Zoológico." *Revista del Jardín Zoológico de Buenos Aires. Época II* 8, no. 33 (April 1913): 3–13.

Onelli, Clemente. "Idiosincrasias individuales de los pensionistas del Jardín Zoológico." *Revista del Jardín Zoológico de Buenos Aires. Época II* 9, no. 34–36 (July–December 1913): 69–76.

Onelli, Clemente. "Idiosincrasias individuales de los pensionistas del Jardín Zoológico." *Revista del Jardín Zoológico de Buenos Aires. Época II* 10, no. 37–38 (July 1914): 1–7.

Onelli, Clemente. "Idiosincrasias individuales de los pensionistas del Jardín Zoológico." *Revista del Jardín Zoológico de Buenos Aires. Época II* 11, no. 41 (May 1915): 3–28.

Onelli, Clemente. "Idiosincrasias individuales de los pensionistas del Jardín Zoológico." *Revista del Jardín Zoológico de Buenos Aires. Época II* 11, no. 44 (December 1915): 330–36.

Onelli, Clemente. "Idiosincrasias individuales de los pensionistas del Jardín Zoológico." *Revista del Jardín Zoológico de Buenos Aires. Época II* 13, no. 49–50 (July 1917): 3–12.

Onelli, Clemente. "Idiosincrasias individuales de los pensionistas del Jardín Zoológico." *Revista del Jardín Zoológico de Buenos Aires. Época II* 14, no. 53 (April 1918): 2–15.

Onelli, Clemente. "Idiosincrasias individuales de los pensionistas del Jardín Zoológico." *Revista del Jardín Zoológico de Buenos Aires. Época II* 14, no. 54–55 (July 1918): 137–42.

Onelli, Clemente. "Idiosincrasias individuales de los pensionistas del Zoológico." *Revista del Jardín Zoológico de Buenos Aires. Época II* 18, no. 66–68 (October 1922): 133–38.

Onelli, Clemente. "Jardín Zoológico del Sur." *Revista del Jardín Zoológico de Buenos Aires. Época II* 17, no. 65 (April 1921): 115–19.

Onelli, Clemente. "La Arcadia Porteña." *Caras y Caretas* 21, no. 1009 (February 2, 1918). https://hdl.handle.net/2027/iau.31858002547671?urlappend=%3Bseq=474%3Bownerid=13510798903548253-492.

Onelli, Clemente. "La elefantita americana." *Revista del Jardín Zoológico de Buenos Aires. Época II* 2, no. 5 (April 1906): 15–21.

Onelli, Clemente. "La enfermedad del orangután y la coprología." *Revista del Jardín Zoológico de Buenos Aires. Época II* 3, no. 12 (December 1907): 334–37.
Onelli, Clemente. "La étnica social hebrea: Conferencia pronunciada el 30 de junio de 1917 en el Centro Juventud Israelita." *Revista del Jardín Zoológico de Buenos Aires. Época II* 13, no. 49–50 (July 1917): 69–84.
Onelli, Clemente. "La fauna de la América Latina." *Caras y Caretas* 16, no. 744 (January 4, 1913). https://hdl.handle.net/2027/iau.31858002550899?urlappend=%3Bseq=123%3Bownerid=13510798903556134-127.
Onelli, Clemente. "La filosofía de la avicultura práctica: Conferencias hebdomadarias en la Universidad Popular de La Boca." *Revista del Jardín Zoológico de Buenos Aires. Época II* 13, no. 49–50 (July 1917): 85–98.
Onelli, Clemente. "La leche de cabra." *Revista del Jardín Zoológico de Buenos Aires. Época II* 6, no. 23 (July 1911): 195–98.
Onelli, Clemente. "La plaga de los gorriones." *Revista del Jardín Zoológico de Buenos Aires. Época II* 11, no. 42–43 (October 1915): 267–71.
Onelli, Clemente. "La princesa y el pingüino." *Revista del Jardín Zoológico de Buenos Aires. Época II* 7, no. 26 (July 1911): 149–51.
Onelli, Clemente. "Las Geórgicas." *Idea Latina* (1924): 5–6.
Onelli, Clemente. "La sucursal sur del Jardín Zoológico: Su objeto." *Revista del Jardín Zoológico de Buenos Aires. Época II* 3, no. 11 (October 1907): 271–76.
Onelli, Clemente. "La superioridad del animal sobre el hombre (Conferencia pronunciada por el director del J Zoológico en la fiesta anual de la Sociedad 'Sarmiento,' protectora de animales)." *Revista del Jardín Zoológico de Buenos Aires. Época II* 10, no. 37–38 (July 1914): 51–69.
Onelli, Clemente. "Los animales en la vida pública y privada (Conferencia)." *Revista del Jardín Zoológico de Buenos Aires. Época II* 11, no. 42–43 (October 1915): 169–90.
Onelli, Clemente. "Los avestruces en libertad y en domesticidad: Sus costumbres, postura de huevos, cría de sus pollitos." *Revista del Jardín Zoológico de Buenos Aires. Época II* 1, no. 2 (July 1905): 150–68.
Onelli, Clemente. "Los derechos del hombre y del animal." *Revista del Jardín Zoológico de Buenos Aires. Época II* 7, no. 27 (October 1911): 223–41.
Onelli, Clemente. "Meditando sobre la ley de Mendel: Un caso raro no previsto por ella." *Revista del Jardín Zoológico de Buenos Aires. Época II* 4, no. 14 (July 1908): 179–88.
Onelli, Clemente. "Necropsia de un orangután." *Revista del Jardín Zoológico de Buenos Aires. Época II* 4, no. 13 (April 1908): 98–99.
Onelli, Clemente. "No confundir el resabio con la tradición." *Revista del Jardín Zoológico de Buenos Aires. Época II* 10, no. 39–40 (December 1914): 175–90.
Onelli, Clemente. "Nuestro Zoo según 'La Nature.'" *Revista del Jardín Zoológico de Buenos Aires. Época II* 8, no. 31 (October 1912): 267–71.
Onelli, Clemente. "Nuevas adquisiciones." *Revista del Jardín Zoológico de Buenos Aires. Época II* 1, no. 3 (October 1905): 255–56.

Onelli, Clemente. "Nuevas adquisiciones." *Revista del Jardín Zoológico de Buenos Aires. Época II* 3, no. 12 (December 1907): 353–60.

Onelli, Clemente. "Plagas agrícolas y paludismo." *Revista del Jardín Zoológico de Buenos Aires. Época II* 8, no. 29 (April 1912): 33–27.

Onelli, Clemente. "Sección médica." *Revista del Jardín Zoológico de Buenos Aires. Época II* 1, no. 1 (April 1905): 87–94.

Onelli, Clemente. "Sed compasivo con los animales." *Revista del Jardín Zoológico de Buenos Aires. Época II* 7, no. 28 (December 1911): 345–48.

Onelli, Clemente. "Sociología-Zoológica." *Caras y Caretas* 15, no. 724 (August 17, 1912). https://hdl.handle.net/2027/uva.x030598349?urlappend=%3Bseq=855%3Bownerid=27021597769308191-857.

Onelli, Clemente. "Soplando en la llama de 'Ideas.'" *Revista del Jardín Zoológico de Buenos Aires. Época II* 14, no. 54–55 (July 1918): 182–93.

Onelli, Clemente. "Su alteza real la chinchilla." *Caras y Caretas* 15, no. 783 (October 19, 1912). https://hdl.handle.net/2027/iau.31858002550873?urlappend=%3Bseq=1027%3Bownerid=13510798903554933-1073.

Onelli, Clemente. "Tambos modelos y cabrerías (Memorándum pedido por el Sr. Intendente Municipal Dr. Joaquin S. de Anchorena)." *Revista del Jardín Zoológico de Buenos Aires. Época II* 7, no. 28 (December 1911): 279–92.

Onelli, Clemente. "Una mirada retrospectiva: El Jardín Zoológico en el 1905." *Revista del Jardín Zoológico de Buenos Aires. Época II* 1, no. 4 (December 1905): 400–405.

Onelli, Clemente. "Una mirada retrospectiva: El Zoo en 1906." *Revista del Jardín Zoológico de Buenos Aires. Época II* 2, no. 8 (December 1906): 309–21.

Onelli, Clemente. "Utilidad de la coprología: Algunas autopsias que lo confirman." *Revista del Jardín Zoológico de Buenos Aires. Época II* 2, no. 8 (December 1906): 278–86.

Onelli, Clemente. "Variantes sobre nuestro Jardín Zoológico." *Revista del Jardín Zoológico de Buenos Aires. Época II* 8, no. 30 (July 1912): 173–77.

Onelli, Clemente. "Vista retrospectiva: El Jardín Zoológico en 1920." *Revista del Jardín Zoológico de Buenos Aires. Época II* 17, no. 65 (April 1921): 111–14.

Onelli, Clemente. "Zoologia útil: Fauna que desaparece, su protección, las alpacas." *Revista del Jardín Zoológico de Buenos Aires. Época II* 1, no. 3 (October 1905): 241–54.

Orofino Púrpora, Roberto. *Por la unidad de Hispano América*. Córdoba: Martín Fierro, 1951.

Osés, Enrique. "La penetración yanqui de la que nadie habla siquiera." *Crisol*, June 25, 1939.

Otovo, Okezi T. *Progressive Mothers, Better Babies: Race, Public Health, and the State in Brazil, 1850–1945*. Austin: University of Texas Press, 2016.

Pavlovsky, Alejandro. "Sección administrativa." *Revista del Jardín Zoológico de Buenos Aires. Época II* 1, no. 3 (October 1905): 313–19.

Pedernera, Sergio, and Estela Pagani, eds. *Sarmiento, espacio y política: El Parque 3 de Febrero*. Buenos Aires: Dirección General Patrimonio e Instituto Histórico de la Ciudad de Buenos Aires, 2010.
Pegoraro, Andrea S. "El uso de motivos indígenas de colecciones del Museo Etnográfico de la UBA en los incios del siglo XX: Actores, actividades y objetos." *Revista del Museo de Antropología* 10, no. 1 (June 2017): 27–36.
Petersen, Mark. *The Southern Cone and the Origins of Pan America, 1888–1933*. Notre Dame, IN: University of Notre Dame Press, 2022.
"Petronio." *La Nación*, January 2, 1908.
"Petronio: El nuevo pensionista." *La Nación*, October 6, 1907.
Pierce, Joseph M. *Argentine Intimacies: Queer Kinship in an Age of Splendor, 1890–1910*. SUNY Series, Genders in the Global South. Albany: State University of New York Press, 2019.
Pita González, Alexandra. *La Unión Latino Americana y el boletín "Renovación": Redes intelectuales y revistas culturales en la década de 1920*. México, DF: El Colegio de México, Centro de Estudios Históricos, 2009.
Pliny. *Natural History*. Translated by H. Rackham. Vol. III: Books 8–11. Loeb Classical Library 330. Cambridge: Harvard University Press, 1938.
Porcher, Jocelyne. "The Relationship Between Workers and Animals in the Pork Industry: A Shared Suffering." *Journal of Agricultural and Environmental Ethics* 24, no. 1 (February 2011): 3–17. https://doi.org/10.1007/s10806-010-9232-z.
Pouillard, Violette. "The Silence and the Fury: Addressing Animal Resistance and Agency through the History of Human-Animal Relationships." In *Animals Matter: Resistance and Transformation in Animal Commodification*, edited by Julien Dugnoille and Elizabeth Vander Meer, 32–55. Leiden: Brill, 2023.
Poy, Lucas. "Ciudadanía, derechos políticos y conciencia de clase: La cuestión de la naturalización de los extranjeros en los orígenes del socialismo argentino." *Diálogos: Revista Electrónica de Historia* 16, no. 2 (December 2015): 191–217.
Poy, Lucas. *El Partido Socialista argentino, 1896–1912: Una historia social y política*. Santiago: Ariadna, 2020. https://ariadnaediciones.cl/images/pdf/partidoSocialAr.pdf.
Poy, Lucas. "'Positive and Scientific': The Influence of Enrico Ferri in Late-Nineteenth Century Argentine Socialism." *Global Intellectual History* (September 2023): 1–19. https://doi.org/10.1080/23801883.2023.2258462.
Poy, Lucas. "Teoría y práctica del 'primer diputado socialista de América': La acción de Alfredo Palacios en el Parlamento argentino y sus repercusiones en la prensa socialista, 1904–1908." *Nuevo mundo mundos nuevos*, 2018. https://doi.org/10.4000/nuevomundo.71852.
Prado, Antonio. "El alsaciano y el orador del Zoo." *PBT* 9, no. 371 (January 6, 1912): 97.
"Primera fiesta del Animal." *El Monitor de Educación Común* 28, no. 425 (May 31, 1908): 291–308.

Quiroga, Horacio. "La tortuga gigante." In *Cuentos de la selva*, 5–15. Buenos Aires: Buenos Aires Cooperativa Editorial Limitada, 1918.

Ralph. "Tatú-Carreta, del Chaco." *Caras y Caretas* 16, no. 778 (August 30, 1913). https://hdl.handle.net/2027/uiuo.ark:/13960/t7gq9jf7j?urlappend=%3Bseq=1223.

Rea, Lauren. *Children's Culture and Citizenship in Argentina: A History of Billiken Magazine (1919–2019)*. York: White Rose University Press, 2023.

Redondo, Manuel. "Noticias." *Caras y Caretas* 12, no. 801 (February 7, 1914). https://hdl.handle.net/2027/uva.x030598367?urlappend=%3Bseq=819%3Bownerid=27021597769305992-823.

Redondo, Manuel. "Sarrasqueta busca un perro." *Caras y Caretas* 12, no. 801 (February 7, 1914). https://hdl.handle.net/2027/iau.31858002550956?urlappend=%3Bseq=295%3Bownerid=13510798903554277-299.

Regulo, A. "Necedades." *La Vanguardia*, September 18, 1906.

Reybel, J. "Clemente Onelli." *Caras y Caretas* 21, no. 1008 (January 26, 1918). https://hdl.handle.net/2027/iau.31858002547671?urlappend=%3Bseq=366%3Bownerid=13510798903548253-384.

Rinland, Jessica Sarah. *Monólogo colectivo*. Produced by Trapecio Cine and Jessica Sarah Rinland, 2024.

Ritvo, Harriet. "The World as Zoo: Acclimatization in the Nineteenth Century." In *The Ark and Beyond: The Evolution of Zoo and Aquarium Conservation*, edited by Ben A. Minteer, Jane Maienschein, and James P. Collins, 27–38. Chicago: University of Chicago Press, 2018.

Rock, David. *Politics in Argentina, 1890–1930: The Rise and Fall of Radicalism*. Cambridge Latin American Studies 19. London: Cambridge University Press, 1975.

Rodríguez, Julia. *Civilizing Argentina: Science, Medicine, and the Modern State*. Chapel Hill: University of North Carolina Press, 2006.

Rodríguez de Romo, Ana Cecilia, and Martha Eugenia Rodríguez Pérez. "Historia de la salud publica en México: Siglos XIX y XX." *História, Ciências, Saúde-Manguinhos* 5, no. 2 (October 1998): 293–310. https://doi.org/10.1590/S0104-59701998000200002.

Rojas, Ricardo. *La restauración nacionalista: Informe sobre educación*. Buenos Aires: Ministerio de Justicia e Instrucción Pública, 1909.

Rolla, David. "¿Cuándo se modernizará nuestro jardín zoológico?" *Mundo Argentino* 21, no. 1066 (June 24, 1931).

Romero, José Luis, and Luis Alberto Romero, eds. *Buenos Aires: Historia de cuatro siglos*. Vol. 2. Buenos Aires: Editorial Abril, 1983.

Rothfels, Nigel. *Savages and Beasts: The Birth of the Modern Zoo*. Baltimore, MD: Johns Hopkins University Press, 2002.

Ruggiero, Kristin. *Modernity in the Flesh: Medicine, Law, and Society in Turn-of-the-Century Argentina*. Redwood City, CA: Stanford University Press, 2004.

Ruhl, Arthur. *The Other Americans: The Cities, the Countries, and Especially the People of South America*. New York: Scribner's Sons, 1910.

Rusiñol, Santiago. *Un viaje al Plata*. Translated by G. Martínez Sierra. Madrid: V. Prieto, 1911.

Sáenz Peña, Roque. "Resurgere." *Caras y Caretas* 16, no. 744 (January 4, 1913). https://hdl.handle.net/2027/iau.31858002550899?urlappend=%3Bseq=113%3Bownerid=13510798903556134-117.

Salvatore, Ricardo D. *Subalternos, derechos y justicia penal: Ensayos de historia social y cultural. Argentina 1829–1940*. Criminología. Barcelona: Gedisa, 2013.

Salvatore, Ricardo Donato, and Carlos Aguirre, eds. *The Birth of the Penitentiary in Latin America: Essays on Criminology, Prison Reform, and Social Control, 1830–1940*. New Interpretations of Latin America Series. Austin: University of Texas Press, Institute of Latin American Studies, 1996.

Sánchez, Norma Isabel. *La higiene y los higienistas en la Argentina: 1880–1943*. Buenos Aires: Sociedad Científica Argentina, 2007.

Sanhueza Lizardi, Rafael. *Un invierno en Buenos Aires*. Paris: Paul Ollendorff, 1914. https://hdl.handle.net/2027/hvd.32044080469224.

Sarmiento Sociedad Protectora de Animales. "29 de abril—Día del Animal." *Billiken* 4, no. 180 (April 30, 1923): 19.

"¡Sayán ha muerto! ¡Viva Sayán!" *La Vanguardia*, March 30, 1909.

Scarfi, Juan Pablo. "La emergencia de un imaginario latinoamericanista y antiestadounidense del orden hemisférico: De la Unión Panamericana a la Unión Latinoamericana (1880–1913)." *Revista Complutense de Historia de América* 39 (January 9, 2014): 81–104. https://doi.org/10.5209/rev_RCHA.2013.v39.42679.

Scarfi, Juan Pablo, and David Sheinin, eds. *The New Pan-Americanism and the Structuring of Inter-American Relations*. Routledge Studies in the History of the Americas. New York: Routledge, Taylor & Francis Group, 2022.

Scarzanella, Eugenia. *Ni gringos ni indios: Inmigración, criminalidad y racismo en la Argentina, 1890–1940*. Translated by Irene Theiner. Bernal, Buenos Aires: Universidad Nacional de Quilmes Ediciones, 2003.

Schávelzon, Daniel, Marina Vasta, and Patricia Corsani. *El pórtico bizantino del Jardín Zoológico de Buenos Aires: Una reflexión sobre nosotros mismos*. Edited by Liliana Barela. Buenos Aires: Dirección General Patrimonio e Instituto Histórico de la Ciudad de Buenos Aires, 2013.

Schulz, R. "Clemente Onelli." *Boletín de la "Sarmiento"* 15, no. 137–138 (December 1924): 2–5.

Sebastiani, Silvia. "A 'Monster with Human Visage': The Orangutan, Savagery, and the Borders of Humanity in the Global Enlightenment." *History of the Human Sciences* 32, no. 4 (October 2019): 80–99.

Ski. "Los niños y los paquidermos." *Caras y Caretas* 15, no. 721 (July 27, 1912). https://hdl.handle.net/2027/uva.x030598349?urlappend=%3Bseq=385%3Bownerid=27021597769308191-387.

"Sociedad Luz: Universidad Popular." *La Vanguardia*, September 5, 1906.

Solberg, Carl. "Immigration and Urban Social Problems in Argentina and Chile, 1890–1914." *Hispanic American Historical Review* 49, no. 2 (May 1969): 215–32.

Soluri, John. *Creatures of Fashion: Animals, Global Markets, and the Transformation of Patagonia*. Flows, Migrations, and Exchanges. Chapel Hill: University of North Carolina Press, 2024.

Suriano, Juan, ed. *La cuestión social en Argentina, 1870–1943*. Buenos Aires: La Colmena, 2000.

Szczygielska, Marianna. "Animal Sex in Public: Warping Time and Sexuality in the Zoo." *Environmental Humanities* 14, no. 3 (November 1, 2022): 641–60. https://doi.org/10.1215/22011919-9962948.

Tacuabé. "Desde un banco del zoo." *La Nación*, September 15, 1906.

Tato, María Inés. "Del crisol de razas a la Argentina desintegrada: Un itinerario de la idea de nación, 1911–1932." *Historia y Política* 17 (June 2007): 153–73.

Trabajos y actas del Congreso Penitenciario Nacional celebrado en Buenos Aires del 4 al 11 de mayo de 1914. Buenos Aires: Talleres Gráficos de la Penitenciaría Nacional, 1914. http://www.bibliotecadigital.gob.ar/files/original/26/2145/trabajos-actas-congreso-penitenciario-nacional_1914-.pdf.

Uddin, Lisa. *Zoo Renewal: White Flight and the Animal Ghetto*. Minneapolis: University of Minnesota Press, 2015.

"Una pesquisa sensacional." *Caras y Caretas* 15, no. 722 (August 3, 1912).

"Un reportaje al orangután 'Petronio.'" *Caras y Caretas* 10, no. 471 (October 12, 1907). https://hdl.handle.net/2027/iau.31858002533986?urlappend=%3Bseq=174%3Bownerid=13510798903556209-178.

"Una riqueza rural que se desdeña." *Anales de la Sociedad Rural Argentina* 38–39 (1904): 41.

"Una visita al Jardín Zoológico: El sentimiento del arte en las fieras de Palermo." *Caras y Caretas* 3, no. 70 (February 3, 1900). https://hdl.handle.net/2027/hvd.32044094186343?urlappend=%3Bseq=749%3Bownerid=27021597764402912-795.

Urich, Silvia. *Los perritos bandidos: la protección de los animales de la Ley Sarmiento a la Ley Perón*, 2nd ed. Temperley, Argentina: Tren en Movimiento, 2015.

Vaccari, Augusto. "Lo que va de la semana: Zoológicas." *Caras y Caretas* 24, no. 1202 (October 15, 1921).

Vallejo, Gustavo. "La hora cero de la eugenesia en la Argentina: Disputas e ideologías en el surgimiento de un campo científico, 1916–1932." *História, Ciências, Saúde-Manguinhos* 25, suppl. 1 (August 2018): 15–32. https://doi.org/10.1590/s0104-59702018000300002.

Vandersommers, Daniel. *Entangled Encounters at the National Zoo: Stories from the Animal Archive*. Environment and Society. Lawrence: University Press of Kansas, 2024.

Vasey, Paul L., and Volker Sommer. "Homosexual Behaviour in Animals: Topics, Hypotheses and Research Trajectories." In *Homosexual Behaviour in Animals:*

An Evolutionary Perspective, edited by Paul L. Vasey and Volker Sommer, 3–42. Cambridge: Cambridge University Press, 2006.

Vasta, Marina. "De 'adorno utilísimo' a 'paraíso de los niños': La especie Jardín Zoológico a través de la evolución del ejemplar en Buenos Aires." *Registros* 13, no. 2 (July 2017): 46–62.

Vasta, Marina. *Viaje pintoresco y excursión científica: El Jardín Zoológico de Buenos Aires, 1888–1924*. Tesis del IAA. Buenos Aires: Universidad de Buenos Aires, Facultad de Arquitectura y Urbanismo, 2018.

Walter, Richard J. *Politics and Urban Growth in Buenos Aires, 1910–1942*. Cambridge Latin American Studies 74. Cambridge: Cambridge University Press, 1993.

Wesolowska, Aleksandra, Bronislawa Pietrzak, Bozena Kociszewska-Najman, Miroslaw Wielgos, Krzysztof Czajkowski, Ewa Wietrak, Katarzyna Karzel, and Maria K. Borszewska-Kornacka. "Barley Malt-Based Composition as a Galactagogue: A Randomized, Controlled Trial in Preterm Mothers." *Ginekologia Polska* 92, no. 2 (February 26, 2021): 118–25. https://doi.org/10.5603/GP.a2020.0107.

Wolfe, Cary. *What Is Posthumanism?* Minneapolis: University of Minnesota Press, 2009.

Yaverovski, Alejandro Martín. "Tres repúblicas barriales y sus universidades: Las Universidades Populares en la ciudad de Buenos Aires en el período de entreguerras, un esquema preliminar." *Historia Caribe* 16, no. 38 (June 2021): 35–80.

Zavattaro, Mario. "Los hombres prácticos." *Caras y Caretas* 11, no. 532 (December 12, 1908). https://hdl.handle.net/2027/coo.31924057400529?urlappend=%3Bseq=1311%3Bownerid=13510798902637714-1345.

Zimmerman, Eduardo. "Racial Ideas and Social Reform: Argentina, 1890–1916." *Hispanic American Historical Review* 72, no. 1 (1992): 23–46.

INDEX

Page numbers in *italic* refer to figures.

Printed and bound by CPI Group (UK) Ltd, Croydon, CR0 4YY
16/09/2026
14963418-0001